# Critique of the
# Theory of Evolution

# Critique of the Theory of Evolution

WALTER FRIEDMAN

Resource *Publications*

An imprint of *Wipf and Stock Publishers*
199 West 8th Avenue • Eugene OR 97401

CRITIQUE OF THE THEORY OF EVOLUTION

ISBN 13: 978-1-55635-175-4

Manufactured in the U.S.A.

*To those who believe that Darwin was a quack.*

# Contents

# Preface

Every book has its purpose, at least the author hopes so. The purpose of this book is to show that the theory of evolution is scientifically incorrect. This book is not a book on metaphysics or religion.

Chapters devoted to criticism of the evolutionary theory do not contain the words *God, Intelligent Design, Messiah*, etc.; one of the purposes of this book is to provide a scientific evaluation of the evolutionary theory without offering any alternative. However, these words do appear in Part III in discussions of legal matters.

What makes this book different from all the other books in this category? Before answering this question, let's take a look at the other books in the field. In general, they could be divided into two major categories:

1) *Books critical of the evolutionary theory written by theologians*

The authors of these books assume that the Bible is correct. Though they may be right, this is a matter of faith and not of scientific truth, so their books completely miss the target.

2) *Books critical of the evolutionary theory written by proponents of Intelligent Design.*

Their brand of criticism could be summed up in the following sentence: modern life-forms are too complex to be thought to evolve from a single source known as the Original Cell.

But the phrase "too complex" lacks precise scientific meaning; therefore, this form of criticism is far off the target.

These two categories have one thing in common—their critics of Darwinism assume that it is imperative to replace the evolutionary theory with some other theory, be that a biblical account of the creation or an account of God as the Intelligent Designer. In reality, replacing the evolutionary theory with another alternative is an unnecessary requirement. As the history of science shows, the vast majority of erroneous scientific theories fell not because they were replaced with new, more sophisticated theories but because they contained either contradictory statements or statements that led to ridiculous conclusions, or both. Still, many people believe, for a variety of reasons, that a valid criticism of a scientific theory should offer an alternative theory that explains the same phenomena. But anyone familiar with the history of science knows that this is not always true. In fact, many theories, such as those of astrology and dialectical materialism, the concept of ether, the theory of transmutation of elements (this one comes from alchemy), etc., were thrown out without replacement—they were so ridiculous that no replacement was necessary.

The purpose of this book is not to replace the theory of evolution with some other theory or theological system, but rather to show that the weakness of the evolutionary theory's arguments disqualifies it from being called a scientific theory at all.

If I decide to write another book as a continuation of this one I will definitely offer an alternative, but for the time being this is not my intention.

# Critique

# 1 : Pseudo-Scientific Methodology

Every natural science uses its own methodology to derive conclusions; experimental data either proves that the conclusions are correct or disproves them. Physics and mathematics have, arguably, the most advanced and fruitful methodologies that have withstood the test of time (although mathematics is not a natural science, its methodology meets all the criteria of scientific methodology). Before discussing the methodology that biologists use, it would be beneficial to take a brief look at the methodology commonly used by physicists and chemists.

The science of physics begins with the set of propositions called postulates, or laws of Nature. Examples of postulates include Newton's laws of motion and Maxwell's equations of electrodynamics. While postulates themselves are not subjected to experimental verification, certain mathematical manipulations are used to derive conclusions from the postulates. These conclusions are then compared to experimental data. If the conclusions are in agreement with the experimental data then the postulates are correct. For example, you can use Newton's laws of motion to predict the motion of your car, or the motion of an airplane, or the motion of the Earth around the Sun.

Mathematicians use a similar approach; however, instead of postulates they rely upon what are called axioms.

Biologists take an entirely different approach—they do not have postulates or axioms because they do not need them. Unlike chemistry or physics, which are predictive sciences, biology is a descriptive science. In other words, biology does not make predictions but rather classifies animate objects into categories or classes. Some examples of classes are as follows: mammals, birds, and reptiles. Mammals, for example, could be classified further into subclasses such as feline, bovine, and ape. Species that belong to the same subclass possess certain similarities—in physical appearance, hunting habits, mating habits, and the like. Biological classifications are very advantageous because they allow compact descriptions of huge numbers of species.

So far so good. But proponents of the theory of evolution went much further by suggesting that common characteristics indicate that

3

members of a subclass have common ancestors down the evolutionary line—ancestors that are now deemed to be extinct. This is the so-called concept of macroevolution that forms one of the stepping stones of the evolutionary theory.

Could this same methodology that was used to arrive at the concept macroevolution be applied in other branches of science? Let us try to use such a methodology in chemistry and see what happens.

As the periodic table of elements demonstrates, there are several groups of elements with similar characteristics; examples of such groups include the lanthanide series, the actinide series, and inert gases. If the macroevolution methodology is applied to the groups of elements of the Periodic Table, one inevitably comes to the conclusion that the elements of a group were branched out of the same element as the result of unknown chemical reactions. This "primal element" was completely used up in the series of reactions, so cannot, therefore, be found in the native state any longer. Of course, any chemist would say that this is nonsense. The concept of "primal elements" is completely unscientific. But evolutionists use this faulty methodology to support the concept of macroevolution.

Though some evolutionists saw the weakness of the original concept of macroevolution, they decided to strengthen it by saying that there is no alternative explanation of the fact that certain species have very close characteristics, such as nearly identical genetic structure. This, however, is not correct. There are at least two alternative explanations: 1) Proponents of the hypothesis of alien intervention believe that the Earth's species were created by extraterrestrial civilization millions of years ago. One might argue that extraterrestrial scientists liked similarities so much that they couldn't resist the urge to create animals that look and behave alike. 2) A very small number of scientists believe that the universe, and the planet Earth in particular, always existed (this theory originated within Hinduism and was somehow taken up in the sciences). If this theory is correct, scientists would never know why there are unexplainable similarities in animal species.

The concept of macroevolution is based on a faulty logical principle according to which the resemblances among objects indicate a common origin. Centuries ago this principle was deemed to be the foundation of the whole body of science, but currently no one, except for the evolutionists, believes in its validity.

Yet another group of evolutionists decided that the evolutionary theory is a postulate; according to their assertion it does not require experimental proof. But a postulate requires the inference to be subjected to experimental evaluation—if the inference agrees with experimental data then the postulate is correct. So far no one has been able to make a single inference based on the evolutionary theory postulate.

Let us take the most general look at the evolutionists' logic without referring to any particular science. Suppose you are shown photographs of two men who look pretty much alike and are told that they belong to the same ethnic group, live in the same city, and were born in the same year. You may be tempted to conclude that these men are twins, which could be true; however, this is not the only possible explanation of uncanny resemblance. It could be some sort of genetic coincidence that two unrelated individuals closely resemble one another. There are plenty examples of non-related "twins."

# 2 : Darwin's Falsification of the Observation Data

Some species have undergone considerable changes in genetic makeup after being subjected to changes in the environment. For example, indiscriminate use of penicillin caused mutations in certain types of bacteria; all of these changes are described in greater detail in medical literature. Biologists call such changes "microevolution" because they occur within a single species. Presumably, extreme cases of microevolution caused macroevolution.

Man-made microevolution definitely exists; but what about naturally occurring cases of microevolution? Do they exist? Darwin claimed that he observed a case of natural microevolution where the wingspan of a certain type of insect was changed due to a change in wind direction. Initially, the wind at the Galápagos Islands was blowing toward the shore, affording equal chance of survival to all members of the species. Then, as Darwin ascertains, undetermined climatic factors caused the wind to change direction for several years in a row, blowing toward the ocean and thus rendering the island more hospitable to those insects with a larger wingspan. Darwin's claim, however, was never verified and the method of measurement that he employed was not disclosed in any of his books.

It would be helpful at this point to present one of many methodologies capable of determining whether an appreciable change in the wingspan has really occurred. As the reader knows, one or two measurements is not enough; a large number of observations is required to determine the average length of the wings before and after a change in the wind direction. What is a reasonably large number of observations? From a statistician's point of view, ten thousand is a very good number. Of course, a larger number would provide even more reliable data, but a number as large as, say, one hundred thousand is not realistic because it would take forever to complete the experiment. Besides, an increase in precision is negligible when extremely large numbers are used.

Now that we have made, in our thought experiment, ten thousand measurements before the change in wind direction and ten thousand after, what would be the next step? We could compute the average lengths of the wings before and after the change in wind direction and compare them. This, however, is not a good idea—a distortion, or something that the statisticians call "bias," can creep into the measurements. Bias could occur because not all coastal areas are equally represented in the sample, or because of a prevalence of one sex over the other, etc. In fact, there are a number of reasons, both known and unknown, that could cause such a bias.

Luckily, though, statisticians know how to eliminate such bias, so results are usually fairly objective. The following is one of the best methods of removing a bias: an integer number is assigned to each of the 10,000 insects with the numbers ranging from 1 to 10,000, then a table of random numbers is used to select insects from the group. We might end up with about 2,000 objects, but it is almost guaranteed that the selection is random. (You could gain more insight into the theory of random numbers if you consult a wonderful book on statistics, *The Advanced Theory of Statistics* by M. Kendall [vol. I is for beginners in the field; it contains all the data you may need on random numbers].)

Suppose that the average length of the wings before the change in wind direction is 2" and 5" after the change. In this case, you can say with absolute certainty that a change in the wind direction resulted in a change in the wingspan. But what if it was 2" before and 2¼" after? The difference is much smaller now, perhaps due to an error of measurement. Then again, the wind might have caused it as in the previous case. You are not so sure anymore. Luckily, there are statistical methods that allow you to determine with 95% certainty whether the difference in mean values is statistically significant or not.

Darwin did not use any statistical method; therefore, the validity of his conclusion is in question. But there is an even more troubling aspect of his story—there was no appreciable change in the wind direction in the region where he conducted observations. There is only one natural phenomenon, El Niño, that can cause long-term changes in wind direction, but geophysical data shows that the weather remained stable in the region where Darwin took his measurements.

There is only one possible explanation—Darwin committed scientific fraud. The history of scientific fraud is as old as science itself, so there is no surprise here.

As it stands now, naturally occurring microevolution is yet to be observed. As for the man-made microevolution, it occurred in an extremely small number of species—less than 1%—so no definitive conclusion about its effect on the commonality of species can be reached. Most likely, a vast majority of the species possesses a very rigid molecular structure that doesn't allow for appreciable changes in molecular arrangements.

There is yet another way of looking at things. Let us try to transport the methodology that proponents of microevolution use to another branch of natural science, this time physics, and see what kind of conclusion it leads to.

This faulty methodology has been known for centuries. It could be described in short as an "assertion that an effect observed in a particular case actually takes place in all cases under investigation."

It is a well-known fact that uranium decays into lead. These elements are situated next to each other in the periodic table. But if the decay process were extrapolated to all other elements, one would come to the conclusion that all elements transmigrate into their right-hand side neighbors. Of course, every physicist would say that this is rubbish. Evidently, the proponents of microevolution use totally unscientific methodology.

To sum it up, proponents of the evolutionary theory use the kind of methodology that no other branch of natural science would ever use, which makes their theory essentially pseudo-scientific crap.

# 3 : Criticism of Two Basic Principles of the Evolutionary Theory

Oftentimes ill-developed theories lead to ridiculous and laughable conclusions. The evolutionary theory is the most ludicrous theory and its conclusions are an affront to common sense.

Two of the basic principles of the evolutionary theory are the principle of man-made selection and the principle of natural selection, which will be the topic of the next chapter.

Did man-made selection really occur? For example, evolutionists believe that the dog is the domesticated wolf. But wolves are known to be wild and no one in recent history has been able to tame a wolf. It's no wonder that not a single circus show features wolves. Was a "primitive man" unwise enough to try to domesticate a wolf at the expense of his flock of sheep? Or perhaps he decided to prove once and for all that the evolutionary theory is correct even though it would put him into financial ruin. But such dedication is not known even among the evolutionists! Besides, there was a much better candidate: the bear. While the bear is slower than the wolf he is still much faster than the sheep, so he could be trained to guard a flock of sheep. Besides, nobody would try to break into a house guarded by a bear. The bear could be tamed fairly easily; that is why they perform in circuses on a regular basis.

According to the evolutionists, a long time ago people domesticated the wild goose. Now his progeny walks in our backyards and, probably, sings a sad song because his wings are too short to fly.

Oh, look at me my darling,
How long I for your embrace,
But my wings are total disgrace
In my dreams I fly like a starling
But evolutionists separated me from thee!

A "primitive man" could have captured a dozen of these wild geese, which is a fairly easy thing to do, but what would be the next step? As the evolutionists suggest, he separated the slow and low-flying birds from the flock, bred them in captivity, and repeated the process several times until he got the geese that cannot fly. There is one minor problem with this theory, however: our man did not have equipment designed to measure the height and speed of a flight! No wonder he was called "primitive." Besides, in order to take measurements he would have to release the birds from captivity. But once they are free, they are gone, so the evolutionists' idea of domestication of the wild geese is gone with them.

What was the name of the cow's ancestor? Whatever its name, this was a wild animal. But wild animals produce milk only when they have offspring—otherwise, there is no one around to release them from the burden (if you don't milk a cow for several days in a row, she'll literally go insane). A "primitive man" might have tried to milk a wild animal just to learn that nothing comes out of those mammary glands.

# 4 : The Case Studies

The principle of natural selection is to the evolutionary theory as gold-digging is to alchemy. Not very many people believe in alchemy in the twenty-first century, but several other outdated theories are still alive today. The best way to get rid of them is to expose their inconsistencies.

What is inconsistent about the principle of natural selection? To start with, it leads to ridiculous conclusions, as the following cases will show.

## Case Study No. 1

Consider the case of the predator XYZ hunting the prey ZYX. Usually, ZYX has higher speed so it can easily outrun XYZ. But XYZ has a higher acceleration rate, which he uses to his advantage. XYZ gets as close to the prey as possible and then attacks by thrusting his body forward with great acceleration. Still, because the distance between the animals is greater than zero, some ZYX species get away while the slowest ones fall victim to the predator. If the principle of natural selection is correct, only the animals with the greater acceleration rate will survive and pass on their traits to future generations. But then some XYZ members will go home hungry and eventually die prematurely because their acceleration rates are not high enough; the deaths of these slower XYZs will pave the way for future generations with ever-increasing acceleration rates. This analysis shows that XYZ and ZYX cause a mutual increase in acceleration rate. But if this were true, today we would see animals running with super-sonic speeds!

The evolutionists might say that there are physical limitations that prohibit a constant increase in acceleration rate.

Let's denote maximum increase in XYZ by $dX$, and in ZYX by $dZ$. Because XYZ and ZYX have very different body structures, $dX$ is never equal to $dZ$.

A) dX > dZ In this case, ZYX disappears from the face of the earth and XYZ becomes herbivorous (the poor guy has to eat something!). However, no one has observed predators becoming vegetarians.

B) dZ > dX Without its natural enemy, the XYZ species undergoes unconstrained growth. However, this phenomenon has never been observed either.

The evolutionists might argue that this is not just XYZ and ZYX; in fact, they may argue, there is a host of other species involved. This is correct, of course, but it doesn't really matter because some species will be slowly but steadily disappearing due to the increase in number of various types of predators, while the others will be steadily increasing in number due to the lack of several other types of predators. At the same time some predators will become vegetarian because their prey is gone. None of this, however, has been observed so far.

## Case Study No. 2

Darwin used the principle of natural selection in an attempt to explain the fact that some types of fish-eating birds have larger beaks than those of their close relatives. According to Darwin, a larger beak allows a bird to catch bigger fish, thus reducing the number of unsuccessful tries. This may be true, but a majority of the birds do migrate, which means that in some lakes or parts of the ocean they encounter an abundance of smaller fish. Then the large beak becomes a liability—because of the law of inertia, it is much harder to move a large beak than a small one, so the number of unsuccessful tries goes up. Overall, neither size gives definitive advantage.

## Case Study No. 3

Certain animals, including the sea lions, are the "Middle Easterners" of the animal world—they practice polygamy. The strongest animals have female harems that come as spoils from fierce battles. Evolutionists see the mating habits of such animals as proof of the principle of natural selection—the strongest animals pass on their characteristics to future generations by spreading their sperm around, so to speak. This assertion is correct, of course, but, as in case of the sea lions, physical strength is not the most desirable characteristic. For the sea lions who have to cover enormous distances in search of food, the most desirable characteristic is endurance. Actually, in the case of the sea lions, physical strength does not matter at all because they can easily overpower the fish that they eat. As every biologist knows, there is no correlation between physical strength and endurance (just take a look at those dried-up legs of the marathon runners!).

## Case Study No. 4

Evolutionists believe that predators pick out the most genetically imperfect prey—sort of "genetic garbage"—while healthier animals produce the offspring. However, there is plenty of evidence to suggest the opposite.

Consider the case of a healthy, vibrant antelope that has a severe case of diarrhea for two days in a row. One more day would be enough for the strong immune system to overcome a malady, but the predator strikes on the second day and the weakened antelope becomes a prey. The diarrhea overrides the principle of natural selection!

Evolutionists would say that this is a very unlikely event.

But what about the cubs who are more likely to become the predator's lunch? Given a chance to grow up, they could become the fastest animals in the herd.

Often the older animals move slower than "genetically defective" ones. The predators might be munching on the older animal while the genetic defects copulate.

During the hunt a herd breaks into smaller groups, the predator might be chasing a group with normal animals while the animals that are not "supposed" to pass on their genes to future generations join the other groups.

A herd could be subdivided into two groups: Group A includes the animals that are temporarily ill, very young animals, older animals, and normal animals that happen to be in a hapless group chased by a predator; Group B includes animals with genetic defects only. Group A comprises about 90% of the herd; Group B comprises about 10%. There is an 81% chance that the predator will attack a healthy animal. The percentage is high enough to declare the principle of natural selection null and void.

In reality, there is no such thing as the principle of natural selection. But without this principle the evolutionary theory is dead.

# 5 : Too Many Obstacles

## "Cradle of Life"

According to the evolutionists, the ocean is the cradle of life because it is the place where the original cell came to be. Creatures big and small inhabited the ocean, then some of them moved ashore in search of a better life. But the shore was empty and there was no vegetation around. Unless we assume that vegetation moved out of the sea and onto land first, we come to the conclusion that the creatures must have returned to the sea with its abundance of food.

It is not hard to imagine how seaweed found itself on a shore as the result of a powerful storm. But seaweed does not have the luxury of moving in and out of the sea; once it is on the soil it has to adapt to new conditions fast, otherwise it withers away. It takes several generations for seaweed, or any other type of vegetation, to adapt to new conditions. If the evolutionary theory is correctly applied, it leads to the erroneous conclusion that the flora and fauna were never able to leave their original sea habitat.

## Two Sexes

Another obstacle that the evolutionary theory faces is the existence of two sexes.

The evolutionary theory stumbles and falls when its proponents fail to explain the fact that there are two sexes in the animal world. All evolutionists agree that the original cell was asexual—it didn't have reproductive organs. As the evolutionists contend, the reproductive organs appeared later when the organisms became more complex.

There are three possible ways that this could have happened: 1) Some organisms could have developed the penis first, then, say, one million years later the vagina evolved. But the penis alone is a useless characteristic and, if the evolutionary theory is correctly interpreted, its uselessness should have made it disappear quite soon. 2) The vagina could have appeared first. (The same logic as in the previous case applies.) 3) The vagina and penis could have appeared simultaneously.

How great is the chance of that? Besides, since they evolved separately, there is no reason why they should be compatible. For example, it could have been a huge penis and a miniature vagina. More importantly, there is no reason why they should be genetically compatible (genetic compatibility is required to produce offspring).

An asexual organism does not need a partner to reproduce itself; it can branch itself into similar organisms at any time. When two sexes are involved, not everyone gets a shot at it. Some members of a species have to wait until the next mating season. This means that asexual species have a better chance of reproduction and subsequent survival of the species, so asexuality is not a characteristic to be lost. Once again, the evolutionary theory leads to a wrong conclusion.

## Not Knowing How Evolution Occurred

The evolutionists say, "we do not know exactly how evolution occurred, but we know that it occurred because it makes sense to us; otherwise it is impossible to explain similarities between certain types of animals."

Let's see what happens if this line of reasoning is used in the other branches of natural science. According to Einstein's theory of relativity, the mass of a moving object increases along with the increase in its speed and becomes infinite when the object reaches the speed of light. All experimental data shows that this is, indeed, the case. Now, imagine Einstein saying, "the objects become lighter as their speed increases. I do not know why this happens and, frankly, I do not care why, but this makes sense to me!" Needless to say, this "Einstein" would never become a household name.

But this is exactly what the evolutionists do—while there is no experimental data to support their claim, they simply state that the evolution hypothesis makes sense to them. Very scientific, isn't it?

## The Platypus Logic

The evolutionists are experiencing great joy—an overly enthusiastic evolutionist discovered the missing link between the sea creatures and the earth creatures. This specimen is called the "missing link" because he has characteristics of both ocean inhabitants and earth inhabitants.

Here is the question—does the fact that a specimen has characteristics of both species A and species B imply that a transition from A to B, or from B to A, took place? According to the evolutionists, this, indeed, is the case; otherwise, they wouldn't be looking for the "missing links."

What about the platypus, then? It has characteristics of both the bird and the mammal—it lays eggs and at the same time produces milk. However, the evolutionists insist that birds and mammals constitute two distinct branches that did not evolve from each other; they just have a common ancestor, whoever that creature may be.

What else is new? Like all idiots, evolutionists make contradictory statements.

What about the mule? It has characteristics of both the horse and the donkey. However, the mule is not a missing link between the horse and the donkey; if anything is missing here it's the evolutionist's brain.

Now, imagine an earth creature taking a bath in shallow waters and in the process getting impregnated by a sea creature. This is how the missing link the evolutionists are so proud of was conceived! The evolutionists will, certainly, say that this is an impossible scenario. Yes, it is; but the "missing link" is an equally improbable scenario.

# 6 : Mutations

Evolutionists use advances in the science of genetics in attempts to justify the evolutionary theory. From the evolutionist's point of view, the most important genetic concept is the concept of germinal mutations (unlike somatic mutations, germinal mutations are transmitted to the succeeding generations).

Germinal mutations could be divided into two categories: 1) Mutations that occur due to known external factors such as radiation, pesticides, herbicides, etc. These mutations occur on a large scale. These mutations are also called "non-random mutations"; the word *non-random* is used to indicate that the cause of a mutation is known. 2) Mutations that occur due to unknown internal factors—these mutations are caused by changes in cellular structure. These mutations are also called "random mutations"; the word *random* is used to indicate that the exact cause of a mutation is unknown. Two or more random mutations cannot occur at the same time because simultaneous occurrence would classify them as non-random mutations.

In an attempt to avoid the search for a cause of mutations that occurred millions of years ago, evolutionists take into consideration random mutations alone. This is a reasonable assumption from a logical standpoint; however, from a mathematical standpoint it causes enormous difficulties.

Evolutionists use the following explanation of how random mutations spread to the entire population:

> Suppose a random mutation beneficial to an organism's survival occurred in a single individual. Also assume that at least two of the descendants of a couple will receive new characteristics; this is a good assumption because a vast majority of the animals produce more than two cubs during their lifetimes. In a short period of time new characteristics will spread to the entire population at an exponentially increasing rate.

However, the evolutionists' calculations are completely wrong because they do not take into account the rate of survival.

Denote by N the total number of members of a population, by q the average number of members that will reach the reproductive age, and by R the survival rate. Then $R = q/N$. Obviously, $0 < R < 1$.

Denote by K the number of generations that will retain newly acquired characteristics that came as the result of a random mutation, by PK the probability that this characteristic will be found in the K-th generation. It can be proven that $Pk = A \times R^K$.

Here, A is a proportionality constant depending on the average number of cubs. (Readers with a basic knowledge of the probability theory could consult Appendix A.)

With each passing generation Pk goes rapidly down and after about 20 generations it is, for all practical purposes, equal to zero.

This is how the probability theory puts the evolutionary theory to shame. It is no wonder that the geneticists and mathematicians are among the most ardent opponents of Darwin's theory.

There is another reason why geneticists are opposed to the evolutionary theory. It has something to do with the conclusions that contradict experimental genetic data.

Recently several groups of geneticists came to the conclusion that the human genotype could be traced to a woman who, probably, lived several thousand years ago somewhere in North Africa; they even dubbed her "Eve." Not all geneticists agree with this conclusion, but even those who disagree do not state that such tracing is impossible in principle; they simply say that it is beyond the level of modern science, but that in the not-so-distant future it will become a reality. This tracing leads to three possibilities:

1) *The human race has two progenitors, a man and a woman.*

For the evolutionists, this opens a can of worms with all kinds of unsettling questions about religion.

2) *The trail goes beyond the human race and stops with two pre-historic beings, say, male and female apes.*

This raises another set of unsettling questions—why does the trail stop here? Who were these beings? Were they created by an extraterrestrial race? What is the meaning of this?

3) *The genetic trail leads all the way back to the original cell.*

This is a sign of even more trouble! For the original cell is a single parent whose existence implies that all living organisms on this planet carry its genetic imprint. By isolating a common denominator, the imprint of the original cell, in primitive organisms, geneticists should be able to determine how inorganic matter evolved into organic matter with a DNA-like structure that produced the original cell. However, as the research in the field of genetics shows, the imprint of the original cell is a fiction that has nothing to do with serious science. Evolutionists ventured into the field of genetics but came up empty-handed.

Now we are going to present the extremely important argument above in a nutshell, so the evolutionists can read it and weep!

In biology textbooks the DNA compounds are described as long strands of molecules bonded together; modern technology allows these long strands to be cut into smaller ones in an arbitrary manner. If the theories of the original cell, or original cells, are correct, it is possible to cut human DNA, or the DNA of any other species, into smaller and smaller strands in such a way that at least one strand will have a structure identical to that of an original cell. If this strand, also called the imprint of an original cell, is placed into a chemical solution containing enough building blocks, it will imitate the function of an original cell by duplicating itself. As all geneticists agree, such duplication has not been observed so far. A small number of geneticists contend that the imprint mutated beyond recognition and that the original structure is gone for good. However, a majority of the geneticists counter their argument by saying that such mutation would have destroyed the self-reproductive capability.

# 7 : The Original Cell as an Entity That Cannot Possibly Exist

Evolutionists believe that an as yet unknown chemical reaction turned inorganic matter into a DNA-like structure called the original cell. This transformation, they say, occurred millions of years ago. So far, however, nobody has been able to reproduce the reaction and there are no theoretical considerations showing that such a transformation is possible. In fact, the exact opposite is true—theoretically, it is impossible for this kind of transformation to occur. The proof is presented in Appendix B of this book. Unfortunately, the proof is extremely complicated and requires considerable knowledge of atomic physics; for this reason it has been placed outside the main body of the text.

There is yet another angle of attack on the concept of the original cell that requires only a moderate, high-school-level knowledge of physics and chemistry.

We'll start with comparing modern geophysical conditions with the conditions that caused the creation of the purported original cell. According to the cosmological theories, Earth was much hotter those days and the original cell's natural habitat, the ocean, was heated almost to the boiling point, with the highest temperature at the seabed where volcanic activity was affecting the earth's crust. Another noticeable difference was high-intensity bombardment of the planet with alpha-particles, gamma-particles, and beta-particles. All these primordial conditions could be easily reproduced in a physics laboratory. In fact, these conditions are present in one form or another in elementary particle accelerators. But physicists also know that alpha-, beta-, and gamma-particles produce unstable configurations (matter) that disintegrate into original components in a fraction of a second. This means that all cosmological theories indicate that the original cell could not possibly exist.

American astrophysicist Carl Sagan tried to circumvent this difficulty by proposing the theory that a spore of unknown vegetation from an unknown place (was it the Planet of the Apes?) somehow got into outer space, then, while being pushed by solar winds, traveled for trillion of years and finally hit the earth, accomplishing what the original cell failed to do.

Sagan knew physics—there's no doubt about that—but he had no knowledge of genetics; otherwise, he would have known that the "Sagan spore," being a single parent, would have left its genetic imprint on all of the earth's organisms (see chapter 6 of this book).

What if it were possible to design an experiment that would prove once and for all that the evolutionary theory is incorrect? Actually, such an experiment *can* be designed—it is based on the notion of an ecosystem. This is how Webster's New World Dictionary defines *ecosystem*: "a system made up of a community of animals, plants and bacteria and its interrelated physical and chemical environment."

What would be the single most important characteristic vital to the survival of a species? That would be the ability to live outside of an ecosystem. Such an ability also implies the ability to live within any partially developed ecosystem. The original cell was all by itself; there was no ecosystem to support it. If the evolutionary theory is correct, all organisms, past and present, should have inherited its ability to live outside the ecosphere.

The idea for this experiment is simple: it would require only a large aquarium filled with seawater and rock formations and a single specimen of fish, bacteria, or seaweed that would normally inhabit such an environment. Can it support itself outside the ecosystem? Other than coral reefs, no one can exist in such an environment for long.

This experiment is the foundation of science; it shows that the evolutionary theory is *anything* but science.

# 8 : Anthropology

Anthropologists are among the staunchest defenders of the evolutionary theory. This comes as no surprise, for their livelihood depends on it! But anthropologists use a methodology that is even worse than the one used by biologists. It starts with a definition of human beings that defies all rules of logic. This definition consists of several parts; we'll go over each of them.

1) *Only human beings can use objects that are not part of their body to accomplish a multitude of tasks.*

Nothing could be further from the truth; for example, crows use sticks and stones to procure food that is otherwise impossible to get. Anyone who likes to watch TV channels that feature wildlife knows how crows operate.

2) *Human beings are the only ones capable of drawing pictures of their surroundings.*

There are pictures of flowers and trees drawn by elephants living in Thailand; some of these pictures are even posted on the Internet.

3) *Animals are afraid of fire; only human beings can sit next to it.*

Actually, this is an urban legend. Recent evidence suggests that certain types of animals, including the fox, are not afraid of fire.

4) *Only humans use fire to cook food.*

This is true, of course, but anthropologists go a step further by suggesting that "human ancestors," including the Cro-Magnon, were using fire to cook their supper. However, forensic experts disagree by saying that it is simply impossible to ascertain what kind of food was used in that era because the time span between then and now is simply too large for us to draw any conclusion.

5) *Humans are the only species who possess
highly evolved linguistic skills.*

But anthropologists confuse two different things; it would be helpful
to sort them out. To start with, each animal species, not just humans,
has its own language—this is a well-known fact. Human languages
are different from animal languages in one aspect—they are capable
of conveying abstract concepts while animal languages deal with
concrete information only. There is strong evidence that prehistoric
beings who left paintings on the walls of their caves did not have a
language capable of dealing with abstract concepts. This means that
prehistoric beings were, in fact, animals and not human ancestors.

6) *Only human beings can manufacture clothes.*

This is correct, of course; however, there is no evidence that prehistoric beings were wearing any clothes. Prehistoric beings are depicted
in biology textbooks wearing some kind of fur clothing, but this is
pure artists' fantasy. It is impossible to tell what kind of garment, if
any, they were wearing because garments get destroyed after enduring
millions of years of harsh conditions. All that is left are the bones of
prehistoric beings and their primitive weapons.

7) *Only humans are bipedal and use their hands for manual labor.*

The kangaroo is bipedal too, and it uses its hands to get its babies in
and out the pouch. There were also reports in the media about an
Australian joker who taught his kangaroo how to fight on a boxing
ring. Besides, it is not clear if the prehistoric beings were bipedal. For
example, some of them had a leg structure very similar to that of the
kangaroo. This finding suggests that they were moving in long jumps.
Usually, anthropologists compare the length of the arm to the length
of the body and if the arms are relatively short, they conclude that
a species is bipedal. But this reasoning is not without a flaw—there
have been reports of humans who were raised by animals and were
able to move on all fours faster than professional sprinters.

    Every science begins with the measurement. Anthropologists
perform all kinds of measurements on the skeletons of prehistoric beings, then form various ratios such as the ratio of foot length to body
length, nose length to head length, hip width to body length, and so
on. After that, the average value of each ratio is computed and com-

pared to its human counterpart. Then anthropologists happily declare that prehistoric beings are, indeed, human ancestors because the differences ratios between prehistoric beings and humans is very small. But what does "very small" really mean? Is it 0.46%? Or 3.7%? Or 10.2%? Or 24.9%? Is it possible to make the following determination: if the difference between the ratios is less than or equal to 12.6% then the prehistoric species in question is a human ancestor; otherwise, it is not? It is highly unlikely that even anthropologists would make such an idiotic assessment. There is no theory or even a hypothesis behind these measurements, which makes them completely useless. The only proof that prehistoric beings are human ancestors would come from establishing that they had human blood types. But skeletal remains give no indication of blood type.

Anthropologists face another insurmountable problem that could be summed up in the question, Was the transition from one type of prehistoric being to another abrupt or gradual? Abrupt transitions are like those horror movies with ugly-looking extraterrestrial babies shooting out of human mothers. Luckily, all geneticists agree that such a transition is impossible. What if the transition was gradual? This would indicate the existence of transitional species, but so far anthropologists have discovered none.

# 9 : Forensic Science

What color was dinosaur skin? Some paleontologists suggest that it was gray while others think that it was colored like a rainbow. This disagreement is not a big deal; paleontologists know that the skeleton offers no clues as to what the skin color might have been.

All prehistoric beings displayed on the evolutionary chart have very hairy bodies, like that of a monkey. In reality, no one knows what their exteriors looked like because the skeleton offers no information regarding the skin's look or constitution. There are several other possibilities: a) their skin looked like a baby's ass, which might not be a nice thing to say; b) their skin was covered with some sort of rudimentary scale; there are even modern-day humans with this type of genetic disorder; or c) their skin was covered with a fur similar to cat fur.

Any of these possibilities spells the doom for the evolutionary theory.

Everyone familiar with forensic science knows how difficult it is to extrapolate the look of facial features from the structure of a scull. Usually, additional information such as age, race, sex, etc., of a subject is required to do a fairly good extrapolation. The skeleton provides, with a certain degree of precision, information about the age and sex, but almost never provides information about the race, so an extraneous source of information is needed to determine this important characteristic. Still, the number of good forensic artists is in the double digits worldwide because this is an extremely difficult undertaking.

No one knows how many races of prehistoric beings of one type or another there were, or what the members of different sexes looked like. There is not enough data to make sexual differentiation possible. There is no way of telling what their age was unless they were in their teens, with a smaller than average body. Yet anthropologists draw pictures of human-like prehistoric beings whose faces express human-like emotions and then say, "look at these people, they are almost like us!"

Emotion-wise, this is a very powerful message, but it is completely wrong. Those beings looked like anything but us. Once emotions are removed from the equation, everyone sees clearly that anthropologists went out on a limb.

# Attack on the Straw Man

# 10 : First-Generation Evolutionists

Evolutionists often complain that, as far as the evolutionary theory is concerned, their opponents created a straw man and now merely attack him. The straw man is here, all right, but he was created by the evolutionists, as the history of the evolutionary theory shows.

Every serious science is based on a philosophical system that helps to interpret its experimental data and improve its methodology. For example, modern-day physics and chemistry are based on a philosophical system called positivism; in the past they were based on the doctrine developed by the German philosopher Immanuel Kant.

Biology is the only science that uses the currently defunct Hegelian philosophical system. Hegelian doctrine was not completely wrong, but it was replaced by more advanced philosophical systems, including positivism. This is a very interesting topic; however, a comprehensive discussion of the Hegelian philosophical system is beyond the scope of this book.

First generation evolutionists were firm believers in the Hegelian philosophical system, which comes as no surprise—at that time it was the most advanced philosophical doctrine. Positivism was only in its nascent state.

According to the main principle of Hegelianism, a process is bound to happen when the conditions favorable to its happening are present. The presence of favorable conditions is in itself a proof that the process is predetermined. Nothing in this definition indicates that the process should happen only once; on the contrary, the process will repeat itself as long as favorable conditions are present. For example, the presence of a gravitational field causes objects situated on the earth's surface to remain there for an indefinite period of time; the process will repeat itself for as long as the earth exists. At a certain distance from the earth the condition known as weightlessness exists; weightlessness is also a persistent process that repeats itself permanently.

If the main principle of Hegelianism is properly applied, it predicts that the process that created the original cell will be repeating itself over and over because the elements known as the "building blocks of life," or the constituents of the original cell, are always present. In other words, scientists should be able to observe the creation of an original cell on a constant basis. As everyone knows, this is not the case.

Naturally, the first generation of evolutionists tried to bypass this difficulty by saying that the presence of the building blocks alone is not enough to bring the original cell to life; something else is also needed to start the process of generation of the first DNA-like organic structure. What is this additional condition? There were several explanations depending on which evolutionists you talk to.

A majority of the first generation evolutionists used the concept of "vital force" in a feeble attempt to explain why there was only one original cell. Webster's New World Dictionary gives following definition of *vitalism*: "the doctrine that the life in living organisms is caused and sustained by a vital force that is distinct from all physical and chemical forces and that life is, in part, self-determining and self-evolving." No one really knows what a vital force is because it is a purely metaphysical quality that cannot be measured in any way; the absence of measurements makes it a nonexistent quality. Still, it would be interesting to know why, as evolutionists believe, it acted only once to produce the original cell. The explanation is just as ridiculous as is the concept of a vital force itself. It states that a primordial vital force gave itself up completely while giving life to the original cell, until there was nothing left!

A competing school of "thought" gave an entirely different explanation: in addition to the building blocks, favorable positioning of the planets in the solar system was necessary to bring the original cell to life. This positioning occurs once every $10^{10}$ years or so. This group of evolutionists stooped to the lowest level scientists could imagine: astrology.

Yet another group of evolutionists came to the conclusion that some sort of "nature magic" that acts only once was involved in all this; that group was also known for riding naked on a broomstick.

Without a doubt, the evolutionist will, upon finishing this chapter, cry—the straw man is down! But it was their predecessors who created the straw man to begin with.

# 11 : Second-Generation Evolutionists

The second generation of evolutionists did not like one bit what the first generation did to their beloved theory, so they came up with ideas of their own. As a result, they were able to discard the ideas of vital force, "nature magic," astrology, and all other unscientific garbage. But in the process they created an equally large mess.

Initially, the second generation stated that the creation of the original cell was a random event and, because of this randomness, cannot be reproduced.

In order to proceed with the discussion, we will need to define the word *random*. This is how Webster's New World Dictionary defines it: "without careful choice, aim, plan, etc.; lacking aim and method." Nothing in this definition suggests that a random event is necessarily a one-time event.

Science indeed deals with random events; for example, the landing of an electron on a laboratory screen is a random event in the sense that determination of the exact landing spot is impossible. However, it is possible to calculate the probability of the electron hitting any given spot based on the position of the screen and the electron gun.

Does any branch of science deal with one-time events that cannot be reproduced? The answer is a categorical NO—all scientific theories deal with events that can be reproduced numerous times, otherwise there would be no difference between science and unscientific speculation.

Evolutionists tried to bypass this methodological difficulty by saying that the big bang was also a one-time event that, nevertheless, forms the basis of all cosmological theories.

Let's take a look at an alternative to the big bang theory. If the big bang never occurred, then it follows that the universe always existed. But the notion of a universe that has no beginning runs contrary to all known cosmological data, such as the expansion of the universe, background radiation, etc. On the other hand, the data that would prove that any alternative to the evolutionary theory is incorrect does not exist.

Some evolutionists tried to remedy the situation by asserting the following proposition: at the present time we do not know what kind of biochemical reaction led to the creation of the original cell; however, in the future, scientists will be able to reproduce it and understand why the reaction is extremely rare. But in this case the evolutionary theory is not really a theory but a hypothesis and should be classified as such. This is how Webster's New World Dictionary defines hypothesis: "an unproven theory, proposition, supposition, etc., tentatively accepted to explain certain facts or provide a basis for further investigation." But evolutionists demand complete and immediate acceptance of their ridiculous theory.

Recently an evolutionist proposed that there was more than one original cell; perhaps they came to be in different geological epochs. Although a vast majority of evolutionists disagree with it, this proposition deserves a close look.

If the proposition is correct, then there are two distinct possibilities: 1) All original cells lead to identical evolutionary lines. This is clearly absurd: if this proposition is correct, then dinosaurs, for example, would still be roaming around. 2) The original cells lead to different evolutionary lines. In this case, the number of original cells is impossible to determine and it is also impossible in principle to make classifications of species based on the evolutionary lines. But true science does not deal with objects that defy classification.

Now it's the right time for the evolutionists to yell—the straw man is on fire!!

But the straw man is a creature of their own making.

# 12 : Third-Generation Evolutionists

Recently, NASA probes landed on one of Saturn's moons, Titan. That was the day when the third generation of evolutionists decided to boldly go where no one has gone before by transporting the evolutionary theory to Titan. They hypothesized that Titan's rivers, which are filled with frozen methane (it's awfully cold out there!), are somehow filled with the "building blocks of life" as the earth's rivers were millions of years ago. Of course, the earth was never that cold, but it didn't matter because the earth's landscape looked exactly like its current-day Titan counterpart (This is not a creationist joke; all Jet Propulsion Laboratory (JPL) scientists interviewed by the journalists used this argument!). Now we all know where the straw man came from!

The third generation in effect sided with the first one by using the outdated Hegelian philosophical system mixed with astrology and black-and-white magic sprinkled with vital force. Will naked flights to the moon follow? Luckily, O'Keafe flew out of NASA pronto.

If anyone is to be credited with causing the most damage to NASA, and to science in general, that would be O'Keafe. On his watch NASA embarked on a journey to nowhere by developing extremely complex and costly equipment intended for the search of signs of bacteria on Mars. But not a single type of bacteria can exist outside the ecosystem, and there is none on Mars. O'Keafe wasted tons of money on useless equipment and on the small army of biologists he hired to participate in designing the equipment. As a result, the other, much more important projects were either put on hold or were underfunded. The most disturbing news was O'Keafe's decision not to repair the Hubble telescope, thus effectively shutting it down. The astronomers and science buffs protested ever so loudly, but O'Keafe ignored their protests and went on a hunt for non-existant bacteria.

It's over now, at least for O'Keafe; and the Hubble telescope will be repaired if the reports are correct. This is extremely good news for the spectators interested in true science. As for the third generation of evolutionists, they acted out of desperation because their pseudo-scientific theory is in the process of being destroyed.

Every time NASA makes a technological achievement its biologists claim that they are one tiny step away from proving that life in the Universe came to be as a result of an unknown biochemical reaction. The latest assertion is that this reaction occurred at the time of the big bang explosion, then some of the materials with a DNA-like structure were stored on comets that later hit the earth.

This theory is so pathetic that only a minimal amount of criticism is needed to bring it to its knees.

Let's assume for the sake of discussion that the theory is correct—a number of comets contain materials with DNA-like structures. Then there is the question, when were these materials created? There are two possible answers:

1) *They were created after the creation of the elements of the periodic table.*

But this leads to a previous discussion (see Appendix B), which shows that it is impossible to assemble DNA-like materials from the elements of the periodic table.

2) *They were created at the same time as were the elements of the periodic table.*

That was the time of intense X-ray radiation. Not a single DNA-like material can survive such extreme conditions without decomposing into its elemental ingredients—all radiation experiments prove it!

There is only one possible conclusion: DNA-like materials that brought life to the earth can survive under *any* condition. But if this were true, there would have been thriving civilizations on Mars, Saturn, Jupiter, etc.—all these planets were hit by the comets. But this is only half of the story—all earthling men and women, being descendants of the original super-cell, would have been able to walk through atomic explosions, jump out of airplanes without a parachute, drink volcanic lava as if it were a soft drink, etc., because they all are carriers of the imprint of the original super-cell. Who wants to be a Superman?

Often scientists working for NASA cite the famous Occam's Razor principle as a "proof" that life somehow came to be by itself. It would be relevant at this point to take a close look at Occam's Razor.

## Occam's Razor

This principle is also called the Law of Parsimony or the Law of Economy. In its original form the principle states the following: "Plurality should not be posited without necessity." In other words, out of two or more competing theories the simplest one should be chosen.

There is nothing wrong with the principle; indeed, all major sciences follow it, or at least try to. But incorrect interpretations are abundant.

The statement is not unconditional; it contains ostensive conditions marked by the word "necessity." Clearly, there are cases where an increase in plurality is unavoidable.

Consider, for example, the application of the principles of quantum mechanics to solid and liquid bodies. These principles alone are not enough. In the case of solid bodies, symmetry considerations reflecting the structure of the crystals are added to the basic principles of quantum mechanics. In the case of a liquid, the principles of quantum mechanics alone do not explain such phenomena as super-fluidity; Landau's theory dealing with the mixture of two liquids with vastly different properties was put forward to explain this phenomenon.

It is impossible to assemble a material with DNA-like structure from the elements of the periodic table (again, see Appendix B), therefore the addition of a broader principle is, in the words of the Occam's Razor, a necessity.

The principle is quite straightforward and easy to use; the possibility of misuse is minimal. But the evolutionists managed to misuse it because they are brain-dead.

# 13 : R. A. Fisher

Evolutionists claim that theoretical and experimental genetic data prove that the evolutionary theory is correct. Usually they site the works of R. A. Fisher. However, geneticists claim the exact opposite.

R. A. Fisher (1890–1962) was an outstanding British mathematician and geneticist. He is one of the two founders of the mathematical theory of statistics. However, none of his works or the works of other geneticists with a background in mathematical statistics deals, directly or indirectly, with the evolutionary theory.

Evolutionists often cite the famous Fisher theorem on the variance of species as a proof of validity of the evolutionary theory. Fisher's variance theorem states the following: "The rate of increase in fitness of any organism at any time is equal to its genetic variance in fitness at that time."

In the context of the Fisher theorem, *variance* is a mathematical term. This is how Webster's New World Dictionary defines the square root of the variance called "deviation": "a measure of the way items are distributed in a frequency distribution."

There is plenty of empirical data that confirms Fisher's theorem; one of the experiments is described in Encyclopedia Britannica.

> This theorem has been confirmed experimentally. One study employed different strains of *Drosophila serrata*, a species of vinegar fly from eastern Australia and New Guinea. Evolution in vinegar flies can be investigated by using "population cages" and finding out how a population changes over many generations. Experimental populations were set up, with the flies living and reproducing in isolated microcosms. Single-strain populations descended from flies collected either in Popondetta, New Guinea, or in Sydney, Australia; and a mixed population was established by crossing these two strains of flies. The mixed population had the greater initial genetic variation, since it was started by combining two different single-strain populations.

Two results deserve notice. First, the mixed population had, at the end of the experiment, more flies than the single-strain populations. Second, and more relevant, the number of flies increased at a faster rate in the mixed population than in the single-strain populations. (Encyclopedia Britannica, CD-ROM 2001)

The Fisher theorem is correct—no doubt about that—but it has nothing to do with the evolutionary theory because it doesn't state that evolution of a species causes changes in a population variance. As the article shows, an increase in fitness may be caused simply by a mixing of the fly populations, not by some kind of evolutionary process.

Actually, Fisher proved something quite opposite to the evolutionary theory.

How large is the probability that a mutation asserts itself in a sexually propagating population? Neglecting selection, R. A. Fisher showed that a once arisen mutation symbolized by a genotype Aa in a population of individuals with the genotype AA has hardly any chance to survive. In order to hand down the allele "a" to its offspring, the Aa individual has to pair with an AA individual. The probability of a loss of the allele "a" is given by Fisher as being $e^{-1}$ per generation (e is the base of the natural system of logarithms). The rate of elimination is described by a Poisson-distribution:

$e^{-1} = 0.368$

The probability that the allele a does still exist after one generation is therefore

$1 - 0.385 = 0.632$

and after another generation:

$e^{-(1-0.368)} = 0.531$

This means that in 90% of all possible cases the allele is extinguished after 15 generations. A chance loss or gain of a non-adaptive allele by a population is called a *genetic drift*. . . .

This example shows that the development of the abundance of forms as we know it in nature cannot be caused by the accumulation of neutral mutations. ("Mutation")

Trying to remedy the situation, E. Mayr introduced the concept of gene flow. According to Mayr, the situation changes drastically when natural selection is introduced.

> An advantageous dominant allele spreads very fast, while recessive alleles have an only low ability to assert themselves even if they provide a selective advantage. The size of population is decisive. The smaller a population, the faster the establishment of a new mutation. All new species are developed from small initial populations. E. Mayr (1942) called them founder populations. ("Mutation")

Unlike Fisher, Mayr was a bad mathematician. He got it all wrong: a) if mutation spreads slowly (Darwin believed that it takes at least 1000 generations to develop a new characteristic), then natural selection does not make much of a difference and newly acquired characteristics will practically disappear after 15 generations; b) in order for a new characteristic to take hold, its time of spread has to be less than 15 generations, which contradicts all known scientific data.

Evolutionary biologists are infamous for manipulating both theoretical and experimental data in attempts to convince the general public that the evolutionary theory is infallible. But anyone with even a perfunctory acquaintance with science knows that their contention is hogwash.

## Mathematical Statistics Misused

Starting from the 1930s, a number of astrophysicists supporting the evolutionary theory tried to assess what they call the "probability of incipience of an original cell on another planet." They came up with different numbers ranging somewhere between $10^{14}$ and $10^{67}$ (perhaps the largest number should go to the Guinness Book of Records). All distinguished mathematicians working in the field of mathematical statistics objected to such misuse of their science. These workers include R. A. Fisher, A. Kolmogoroff, E. S. Pearson, J. Neyman, M. G. Kendall, and many others. Their objections are based on the fact that only events that could be repeated an indefinite number of times form the basis of the probability theory (the discharge of electrons from an electron gun would be an example of such events). This example shows that the evolutionists are completely unaware of how the probability theory works. Yet they base their "proof" of validity of the evolutionary theory on the probability theory!

## Disputed Numbers

This is another lie propelled by the proponents of the theory of evolution: there is a consensus in the scientific community that the theory is correct. Nothing could be further from the truth. Not a single geneticist supports it; less than 10% of physicists agree with it, and those who agree have very little knowledge of atomic physics; paleontologists and biochemists are evenly split; and no one knows how many biologists disagree with it because they do not speak their minds for fear of losing their jobs. The only group of specialists who support it wholeheartedly are the anthropologists.

Much support for the evolutionary theory comes from *Scientific American*. But the magazine earned an awful reputation because of unqualified support of another unscientific theory, the so-called Superstring Theory, which defies any experimental verification. Less than 1% of physicists believe that there is any truth to the story of the Superstrings. Perhaps *Scientific American* should change its name to *Idiot's Guide to Science*.

# 14 : Hobbits

Recently, the skeletons of "small people" or "hobbits" who stand only 90 cm (3 feet) tall were discovered on a remote Indonesian island; anthropologists believe that the hobbits and humans have a common ancestor.

The group of archeologists who discovered these small people were interviewed on *60 Minutes*; they say that the first unearthed skeleton was that of a woman. It seems these evolutionists were using circular logic: these beings are humans → this skeleton more closely resembles the human female skeleton → this is a hobbit female → these beings are humans. When this type of logical deduction is used, the first step in the chain of inferences coincides with the last one. In reality, no one knows how to distinguish a female hobbit skeleton from a male one because not a single hobbit has been captured alive or dug out of a fresh grave.

Not a single physicist or chemist uses this type of logical deduction; only the evolutionists are stupid enough to put it to use.

Actually, the very existence of small people proves that the evolutionary theory is incorrect.

What kind of advantage could such small size possibly have? A little body requires a little amount of food—clearly, this is an important advantage. But this is the only advantage. Now, let's take a look at the disadvantages.

1) Attacks by birds of prey. Often, predatory birds attack little children but almost never attack human adults because of their size. The hobbits were as small as human children, so they would have been in constant danger of being attacked.

2) Small people can't hunt in places with tall grass because their vision is blocked by it.

3) Small people can't cover large distances in search of food because their leg muscles are too weak to carry them for prolonged periods of time.

Clearly, the number of disadvantages is larger than the number of advantages. If the principle of natural selection is correctly applied, it proves that these small people are fiction.

Anthropologists are notorious for the bending and twisting of data to fit it into their wacky theories. One of them told *20/20* that the seven year itch is, in fact, a four year itch—it takes four years for an animal couple to raise cubs, and after that the partnership is kaput. Apparently, she had statistical data showing that the number of divorces peaks after four years of marriage, so, in her view, humans follow the same pattern as their animal cousins. There is a problem with this assessment, though—while it takes four years for certain animal species to raise offspring, a vast majority of animals stay with their progeny for only two years, so the whole theory of human-animal behavioral resemblance breaks down.

Some anthropologists decided to look for an explanation of certain characteristics of human behavior beyond the animal world because the animal socialization, in their words, is too primitive. Ants, for example, have a much more complex societal structure with the queen, workers, soldiers, nannies, etc., so these anthropologists stick their heads into anthills in an attempt to find extremely valuable data about human society.

For certain governments, the discovery of "human ancestors" became a matter of national prestige. They lavishly finance anthropological digs and hail their anthropologists as if they were national heroes. Ideally, all departments of anthropology would receive no support from the government; this measure would force the universities to transfer "old bones of human ancestors" to the departments of paleontology.

# SECTION III
# Legal Matters

# 15 : Anti-Evolution Stickers

Some time ago, an Atlanta court ordered the removal of creation-ist stickers from biology textbooks—apparently it was a violation of the principle of separation between state and church. However, it is possible to design stickers in such a way that they would criticize the evolutionary theory without violating the aforementioned principle.

The following could legally be included on such a sticker: "The evolutionary theory is deemed totally unscientific by many scientists. For more information read the articles . . . [list of articles]." There are plenty of articles expressing strong criticism of the evolutionary theory that do not mention any religion; these articles are written by geneti-cists, paleontologists, biochemists, etc., who criticize certain aspects of the theory without offering any religious alternative. Any lawyer would agree that this approach does not go against the principle of separation of church and state; a court order demanding removal of this sticker would violate the principle of free speech. School libraries could be ordered by the Board of Education to have copies of the articles so it would be easier for the students to gain access to them.

Another type of legitimate sticker would be something like this one: "Darwin committed scientific fraud in order to promote racist views reflected in the evolutionary theory. Examples of his racist views are as follows: . . ." (Darwin was never shy about his extremely racist views; there are plenty of examples).

The aforementioned stickers could be used simultaneously—there is nothing in the laws that would prevent such usage.

How about this one: "The evolutionary theory is not the only theory claiming to have the answer to the question about the origin of life. There are other theories as well." The evolutionists' lawyers will argue that this statement refers solely to the theory of Intelligent Design, but they are wrong; there are at least two more competing theories. These theories are as follows: 1) the theory of alien interven-tion stating that extraterrestrial civilization created the earth life forms and 2) the theory, believed by a very small number of scientists, that the universe and its life forms were never created because they always existed.

Neither of these two theories violates the principle of separation between the state and the church.

"As the history of mankind shows, mankind produced several theories regarding the origin of life; these theories include Intelligent Design, the evolutionary theory, the theory of the extraterrestrial origin of earth's life forms, etc." This sticker is simply a historic reference; there is nothing in the US Constitution, or the constitution of any other country, that would prevent the use of historic references in a textbook; otherwise, any reference to a religion should be banished from history textbooks (say goodbye to Zeus and Aphrodite!).

# 16 : Darwin as a Racist

> The break between man and his nearest allies will then be wider, for it will intervene between man in a more civilized state, as we may hope, even than the Caucasian, and some ape as low as baboon, instead of as now between the Negro or Australian and the gorilla. (Darwin, *Descent*, 178)

Up until now, creationists have tried in several states to convince the courts that the evolutionary theory should be removed from biology books because it is extremely racist, but they have failed. Evolutionists brought their lawyers who successfully argued that it doesn't matter whether the evolutionary theory is racist or not because it is correct.

Now any unprejudiced person can see that the evolutionary theory is bullshit; nothing could be more unscientific than it.

Now is a good time to try again; however, a more successful tactic is needed to achieve the removal of the evolutionary theory from biology textbooks. Instead of taking the matter directly to the court, creationists should provide all the pertinent data showing the evolutionary theory's inadequacies and Darwin's racist views to the parents of minority students and ask them to sign petitions demanding withdrawal of the theory from school curriculums. Parents should also be encouraged to instruct their children not to attend biology classes unless the evolutionary theory is removed from their school curriculum. It would be very helpful if local TV and radio stations interviewed parents of the children who choose not to attend the biology classes.

After that, either the parents or the Board of Education will ask the courts to intervene. Hopefully this grassroots campaign will take hold in every state, including the blue states.

> "At some future period, not very distant as measured by centuries, the civilized races of man will almost certainly exterminate and replace savage races throughout the world." (Darwin, *Descent*, 178)

47

It would be fair to say that Hitler heard this call to exterminate lowly races.

The evolutionary theory sets legal precedent—now the KKK and Aryan Race Group can go to court and demand the inclusion of their unscientific theories in school curriculums. The difference between the KKK's ideology and Darwinism is insignificant; this makes the inclusion of KKK literature in biology textbooks very likely. Of course, this predicament could be easily avoided if the evolutionary theory is purged from biology textbooks.

It doesn't seem right to force African American students to study a pseudo-scientific theory that puts them on the same level as the gorilla and calls for their extermination.

Evolutionist lawyers will argue that Darwinism is different from the modern version of the evolutionary theory.

According to legal standards, two scientific theories are deemed to be different from one another if they differ in any respect; otherwise, they amount to the same theory.

Darwinism and the modern evolutionary theory have the same basic principles: 1) the principle of natural selection, 2) the concept of the original cell, 3) the concept of microevolution, and 4) the concept of macroevolution.

They differ in one respect only: the concept of germinal mutation is reflected in modern evolutionary theory, while it was not reflected in Darwinism because it was unknown at that time. But the concept of mutation does not amount to a principal disagreement between Darwinism and modern evolutionary theory. In fact, modern evolutionary theory should be viewed as an expanded version of Darwinism because both theories lead to the same conclusions.

The evolutionists try to separate their theory from Darwin's racist "findings," but this is a futile undertaking.

Evolutionists use the same old argument in courts all across the world: the evolutionary process is not happening now because it could only happen under specific conditions. Apparently, these conditions are not present at the moment.

However, they always fail to specify under which conditions the purported evolution of species occurred. They always say the same thing: we do not know what these conditions were, but they definitely existed in the past.

The best line of attack would be asking the following question: will the conditions under which the purported evolution occurred present themselves in the future?

There are 3 possible answers:

*1) The conditions won't repeat themselves in the future.*
   *It was a one-time occurrence.*

The best comment to counter this answer is this: if you do not know what the conditions are, how can you predict that they will not occur in the future?

There is no way to respond intelligently to this question.

*2) The conditions will repeat themselves in the future.*

This is a modified version of the above comment. If you do not know what the conditions are, how can you guarantee that they will occur in the future?

No intelligent answer is possible.

*3) We do not know whether the conditions will*
   *present themselves in the future.*

This is the rebuttal speech:

All predictive sciences make predictions regarding future and past events. For example, Newton's mechanics makes predictions about events that occurred before and after the measurement of coordinates, velocities, etc. Descriptive sciences are different in this respect because they do not make predictions regarding future and past events; they simply relate objects to categories.

By stating that at the present time there is no way of telling whether or not the conditions that cause the evolution of species will occur in the future, you define the evolutionary theory as a descriptive science. But, being a descriptive science, it doesn't provide any means as to verify that these conditions occurred in the past.

# 17 : Court Battles

In the past, creationists lost practically all court cases to evolutionists because of poor presentations. The purpose of this chapter is to demonstrate a successful strategy. This strategy involves several steps.

Step 1: The best way to put the opponent on defensive is to show right from the start that the basic principles of his theory are false. The first two chapters of this book are devoted to criticism of the basic principles of the evolutionary theory, which are microevolution and macroevolution.

Step 2: This step is a must if the opponent brings anthropologists to testify on his behalf. Otherwise, this step is optional.

Anthropologists' testimonies could be very powerful because they stir strong emotions by showing pictures or drawings of prehistoric beings that look almost like humans. Of course, this is baloney, but the jury and the judges do not know that. A good presentation done by forensic experts is needed to convince the court otherwise. It is very important to take the emotional component out of the equation at the earliest possible stage.

Step 3: Now this is the right time to attack the principle of natural selection. The most comprehensive criticism is presented in this book (see chapter 4 of this book).

Step 4: The attack on man-made selection comes next. This book provides enough material for a successful attack (see chapter 3 of this book).

Steps 3 and 4 are interchangeable, it doesn't matter which one comes first.

Step 5: If the evolutionary theory is correctly applied, it shows that both flora and fauna should have remained in the ocean (see chapter 5 of this book).

Step 6: From the viewpoint of genetics, the evolutionary theory leads to erroneous conclusions (see chapter 6 of this book).

Step 7: Sagan's stupid mistake did not come out of nowhere. It would be very helpful to show the considerations that have led to it (see chapter 7 of this book).

Step 8: The theory of probability shows that random mutations do not spread to the entire population (see chapter 6 of this book). This presentation could be made either by a mathematician or a geneticist with a strong background in the theory of probability.

Step 9: This step could be very tricky, but if properly executed, the victory is almost guaranteed. However, it also has a potential for ruining the case. The material for this step is presented in Appendix B of this book. Although a single person can make good presentation, most likely a team of experts and a coordinator would be called to do the work. These experts include specialists with knowledge of atomic physics, philosophy of science, and, possibly, the basics of the theory of information.

Step 10: This step is almost as good as step 9 and there is no risk in taking it. This step is based on the concept of ecosystems (see chapter 7 of this book).

Step 11: Now is a good time to bring in an emotional component—that would be Darwin's racist views.

Step 12: A strong closing is needed to bring the case home. That would involve exposure of Darwin's fraud (see chapter 2 of this book).

# 18 : History and World Religions

Inclusion of the theory of Intelligent Design in biology textbooks is seen as controversial by many. This doesn't mean that Intelligent Design does not belong in the classroom. A much better approach, from a legal standpoint, would be the inclusion of Judeo-Christian ideas in history textbooks (the same rationale is applicable to those countries with a predominantly Buddhist population where Buddhist ideals could be presented in history classes).

World religions are a part of world history. It is impossible to write a decent book on world history without discussing its major religions; certainly Christianity is one of them. Any court order that prohibits discussion of Christianity or Judaism would, certainly, imply that ancient Greek and Egyptian religions should be banished from history textbooks too, which is clearly a ridiculous proposition.

Usually, the Bible is treated as a religious document; however, nothing prevents schoolteachers from considering it a historic document. In fact, this is how the Bible is approached by Harvard scholars who quote it on a regular basis. There are no limitations on the number of quotations from historic documents, including the Bible; in fact, any limitation would seem arbitrary.

There are several ways of approaching the subject. It seems that the best way would be a discussion of topics pertinent not only to Christianity but to several other religions.

## 1) *The Topic of the Creator*

According to the Bible, there is only one Creator, as the passages show. Many, although not all, Buddhist denominations believe in the entity called Adu Buddha, who created the universe and the life in it; in this sense Adu Buddha is the equivalent of the Judeo-Christian Creator. There are passages in Buddhist texts that refer to Adu Buddha; these passages could be included in a history textbook along with the passages from the Bible and other religious sources. There are many gods in Hinduism, but only one of them, Brahma, is the Creator. Brahma is described in Veda, which is a sort of Hindu bible.

## 2) *Creation of the Universe*

Six stages of creation erroneously called six days are described in the Bible. In Buddhist texts, emphasis is put not on the process of creation but on the structure of the universe, which is its atomic structure. Descriptions are abundant. There are several references to the process of creation in Hindu books.

## 3) *Creation of the Man*

There are passages in the Bible as well as in Veda that cover this topic.

## 4) *Salvation*

Christians and Jews believe in heaven and hell; Buddhists and Hindus also believe in heaven and hell. In addition, these two groups believe in reincarnation.

## 5) *The Savior*

Christians believe that Jesus is the Messiah. Buddhists believe in the Savior-to-be whose title is Maitreya. There is a connection between Jesus and Maitreya, reflected in certain Buddhist texts. Maitreya's Eternal City is called Jerusalem.

The following is from Sanskrit sources:

> "Jerusalem, my happy home
> When shall I come to thee?
> When shall my sorrows have an end?
> Thy joys when shall I see?" (Conze, *Buddhist Scriptures*, 237)

# Evolution, Communism, Psychoanalysis, and All That Jazz

# 19 : The Theory of Evolution
## in Soviet Schools

Westerners tend to think that people who grew up under a Communist regime were heavily indoctrinated with the Communist ideology. This is not quite correct. Indoctrination was very minimal because the officials responsible for Communist propaganda were lazy; they chose the easiest option.

The author of this book had a privilege to spend his school years in a currently defunct USSR secondary school where the level of propaganda was very weak for a variety of reasons. The book that started it all, *The Capital* by K. Marx, was seen to be so sophisticated by the authorities that even college students majoring in physics, engineering, mathematics, biology, journalism, etc., were not required to study it; only the students majoring in economics were forced to go through the nightmare.

Lenin wrote a thin book, *Materialism and Empirio-criticism*, which was not in school or college curriculums either. Lenin also wrote plenty of newspaper articles—about eighty volumes of them—and out of that bunch only two articles, "The April Thesis" and "Three main parts of Marxism," were included in school curriculums. The rest was up to the schoolteachers, who didn't want to burden themselves.

The reader may ask, what were the pupils doing at the Communist studies? Luckily, the Soviet Communist Party held plenty of conferences and symposiums; pupils were forced to memorize the dates of those meetings and brief descriptions of the agendas. All agendas were alike; if you knew one of them you knew them all. The dates were another story. There were plenty of them and pupils with good mnemonic capabilities were getting an unfair advantage. Yet another group of pupils, including this author, was using chi-chi to get by. Chi-chi is a great equalizer; otherwise, you might end up writing, as one of the author's friends did, that Lenin was born in 1917.

But the Communist Party had the single most important propaganda tool at its disposal—the evolutionary theory. The implication was clear—there is no afterlife; you turn into dust after death; love not God but your Communist leaders, especially Lenin; roll up your sleeves and start building Communist society!

Because of an unnatural love of Lenin male homosexuality was rampant. It was reaching 20% of the male population while in the West it was less than 2%.

Ironically, the evolutionary theory was one of the major factors that brought the Soviet version of Socialist society down: it prevented Soviet leaders from introducing economic changes into a faltering Socialist system. You see, according to the evolutionary theory, rich people are the most evolutionarily advanced individuals because of their money-making skills; at least this is how Soviet leaders interpreted Darwinism.

# 20 : The Theory of Evolution and Psychoanalysis

What would happen if the evolutionary theory were proven wrong once and for all? Obviously, anthropologists would lose their jobs because there is nothing for them in a perfectly scientific world (biologists will survive because much of their work does not involve the evolutionary theory). But anthropologists are a small group of pseudo-scientists; their departure will not produce any noticeable effect on the society. However, there is another group of pseudo-scientists whose livelihood depends on Darwin's theory, and this group racks up hundreds of millions of US dollars every year. This pseudo-scientific group believes that any religion is a form of psychosis. Their ideas are based on Freud's psychoanalysis. Since the evolutionary theory is bound to fall and drag psychoanalysis down with it after publication of this book, it would be beneficial to present in this closing chapter a brief but destructive criticism of Freud's invention.

The notion of suppressed, or latent, memories is the foundation of the theory of psychoanalysis; without this concept the theory falls apart.

Rather than analyzing the origin of suppressed memories, it would be more pertinent to ask whether the suppressed memories exist. Acceptance of the concept of suppressed memories leads to insurmountable contradiction, as the following analysis will show.

It is a well-known and easily observed fact that it is impossible to separate one thought process from another, or to stop the train of though, so to speak. This means that the thought process is non-analyzable in the sense that it is impossible to break it into two parts with one part being the thought process itself and the other part being that of an observer who does not affect the part under observation. By the same token, it is impossible to break the thought process into three parts in the following fashion: 1) the thought process itself, 2) the observer part of the thought process that selects certain memories that an individual desires to suppress, and 3) the part of the thought process that suppresses the undesired or harmful memories.

All three parts are the same thought process; there are no lines of demarcation between them. When the mind tries to suppress, consciously or not, an undesired event, say, a car accident, it isolates an event and then attempts to suppress it. The isolation could be achieved only if the event is being constantly held in the memory; otherwise, there would be no distinction between the event under suppression and any other event. For an event to be suppressed, it needs to be taken out of the thought process altogether; otherwise, no suppression is possible. But the thought process is inseparable, as the analysis shows. The observer part that selects undesirable memories and the suppressive part that erases memories do not exist independently of one another. This means that the mind is trying to accomplish two contradictory tasks—it tries to forget an event while keeping it constantly in the thought stream.

One might ask, what about amnesia? Isn't amnesia a proof that suppressed memories exist? But amnesia is not caused by a deliberate attempt to suppress undesired memories; it is caused by external factors such as a blow to the head, electric shock, lack of oxygen, etc. In fact, amnesia shows that only external events can cause memory loss; there is no internal mechanism that would allow one to achieve a similar result.

Do men possess feminine characteristics and women possess masculine characteristics? According to the proponents of psychoanalysis, this, indeed, is the case. But if this is true, then classification of human beings based on sex is an exercise in futility—there is no distinction between the classes of the objects. Every science begins with classification, but in the case of psychoanalysis there is no classification, which means that any inference is baseless. In reality, there are three types of characteristics regarding the sexes: 1) feminine characteristics, 2) masculine characteristics, and 3) neutral characteristics that both sexes have. An example of this third type of characteristic would be a sense of humor.

An inability to recognize the existence of these three characteristics led to incorrect conclusions regarding the causes of homosexuality. According to Freud, male homosexuality is caused by the son's hidden desire to take the place of his mother. But if this assertion were true, there would be no male homosexuals among orphans who were placed in orphanages at a very young age. However, all statistical data proves otherwise.

# Comprehensive Analysis of the Evolutionary Theory

# 21 : Opening Round of Criticism of Darwinism

Over the course of the last decade, evolutionists have developed unusual evasiveness: no matter how you criticize the evolutionary theory, they say that your criticism is misplaced. The best way to beat this charge is to deliver direct criticism of the most prominent books on evolution.

> It seems clear that organic beings must be exposed during several generations to new conditions to cause any great amount of variation; and that, when the organisation has once begun to vary, it generally continues varying for many generations. (Darwin, *Origin*, 31)

Let's take a look at a different statement: the creator, whoever it might be, made organic beings suitable for a great variety of conditions, so there is no need for the organization to vary.

Now, we have a huge variety of animals, some of which are similar and others of which are dissimilar, and two propositions referring to them—Darwin's proposition and mine—and no way of determining which one of them is correct. But the mere presence of two or more conflicting, unprovable propositions shows that the evolutionary theory is not fact but a hypothesis at most.

The division of objects into categories, or classes, is an extremely important scientific task. It begins even before a theory is put forward. To some degree categories are arbitrary; however, there is a general rule that states that objects belonging to one category cannot be put into any other category. For example, the categories of "atom" and "molecule" are two distinct categories, and it is impermissible to interchange them. (Some might argue that it is possible to have matter in solid and liquid forms simultaneously, so categories, in this case the categories of "solid" and "liquid," can indeed mix. But this is incorrect because the solid-liquid would actually be grouped within a new "solid-liquid" category, which is different from both the solid and liquid categories.) There is the reason why categories do not mix: it is impossible to build a valid scientific theory based on mixed categories because improper categorization leads to improper conclusions. But the evolutionists believe otherwise; let's see how Darwin defines the category he calls "the nature of the organism."

> [T]here are two factors: namely, the nature of the organism, and the nature of the conditions. The former seems to be much the more important; for nearly similar variations sometimes arise under, as far as we can judge, dissimilar conditions; and, on the other hand, dissimilar variations arise under conditions which appear to be nearly uniform. The effects on the offspring are either definite or indefinite. (Darwin, *Origin*, 32)

This category is incorrectly defined because it allows similar variations under dissimilar conditions and at the same time dissimilar variations under similar conditions. This categorization is completely wrong because it leads to indefinite effects on the offspring, as Darwin acknowledges; it renders the evolutionary theory unworkable.

The most important problem that every science encounters is the task of uncovering cause-and-effect relationships. Without knowledge of them a science cannot possibly exist. Oftentimes cause-and-effect relationships are incorrectly determined, as the following example shows: Einstein was told by a critic of his revolutionary theory that it is possible to exceed the speed of light if one shines beams of light on a rotating mirror and watches their reflections on a remote screen following one another with a speed that exceeds the speed of light. As Einstein correctly pointed out, this is not a sequence of cause-and-effect events. In any case, failure to indicate at least a probable cause leads to useless theories. Apparently, Darwin did not know about this basic rule of science:

> Each of the endless variations which we see in the plumage of our fowls must have had some efficient cause; and if the same cause were to act uniformly during a long series of generations on many individuals, all probably would be modified in the same manner. (Darwin, *Origin*, 32)

Notice that he writes "probably," not "certainly." Stating that there is a cause without actually showing it is akin to saying, "Tomorrow I will die for an unknown reason. I do not know what the reason is, but I'm certain that death is around the corner." None of this makes any sense, just like the nonsensical theory of evolution.

## Failure of Deduction

> With animals the increased use or disuse of parts has had a more marked influence; thus I find in the domestic duck that the bones of the wing weigh less and the bones of the leg more, in proportion to the whole skeleton, than do the same bones in the wild-duck, and this change may be solely attributed to the domestic duck flying much less, and walking more, than its wild parents. (Darwin, *Origin*, 34)

Does the domestic duck walk more than the wild one? Domestic ducks have their food put in front of them, but their wild cousins have to walk and swim large distances in search of food (there is no food in the air!). If Darwin's logic is correctly applied, the wild duck should have heavier leg bones than the domestic duck.

> I will here only allude to what may be called correlated variation. Important changes in the embryo or larva will probably entail changes in the mature animal. (Darwin, *Origin*, 34)

There is only one kind of change that an embryo or larva can undergo—that would be a mechanical change due to some sort of prenatal or birth trauma. But this kind of change leads to an improperly functioning organism and it is not transmitted to the next generation. Apparently, Darwin has no idea what he's talking about.

> Some instances of correlation are quite whimsical: thus cats which are entirely white and have blue eyes are generally deaf; but it has been lately stated by Mr. Tait that this is confined to the males. (Darwin, *Origin*, 35)

This is quite whimsical—that evolution is affected by the animal's sex. If this were true, there would have been, for a given species, a numerical predominance of one sex over the other. But nothing of that sort occurs in nature.

> The results of the various, unknown, or but dimly under-
> stood laws of variation are infinitely complex and diversi-
> fied. (Darwin, *Origin*, 35)

Everything is in a haze . . . Is it possible to tell that laws exist
without knowing them? No. It is not possible unless one is willing to
introduce into his theory gross violations of logic as Darwin does.

> [T]he domestic races of many animals and plants have been
> ranked by some competent judges as the descendants of
> aboriginally distinct species, and by other competent judges
> as mere varieties. (Darwin, *Origin*, 38)

In other words, any meaningful classification is impossible. But
every science begins with classification; the absence of classification
prevents any fruitful development of a theory.

> In attempting to estimate the amount of structural differ-
> ence between allied domestic races, we are soon involved in
> doubt, from not knowing whether they are descended from
> one or several parent species. (Darwin, *Origin*, 38)

Maybe they are not descended from any species. This statement
by Darwin makes the evolutionary theory look like a guessing game
that lacks even the smallest amount of certainty. Darwin summed
up better than anyone his theory's impotence by saying, "The origin
of most of our domestic species will probably forever remain vague"
(Darwin, *Origin*, 72).

## Vague Ideas

> Nor shall I here discuss the various definitions which have
> been given of the term species. No one definition has satis-
> fied all naturalists; yet every naturalist knows vaguely what
> he means when he speaks of a species. Generally the term
> includes the unknown element of a distant act of creation.
> The term 'variety' is almost equally difficult to define; but
> here community of descent is almost universally implied,
> though it can rarely be proved. (Darwin, *Origin*, 58)

Every science deals with precisely defined concepts. This is the only
way to make a theory work; otherwise any scientific deduction is just

wishful thinking. For example, it is impossible to develop an electromagnetic theory without a proper definition of electric charge. Biologists are the only group of scientists who deal with definitions so vague that they make a theory practically useless.

> Almost every part of every organic being is so beautifully related to its complex conditions of life that it seems as improbable that any part should have been suddenly produced perfect, as that a complex machine should have been invented by man in a perfect state. (Darwin, *Origin*, 58)

There have been plenty of inventions that came out right with the first attempt—the steam engine controller, which was the first feedback control mechanism, is one of them. This idea of beauty is quite poetic, but it never enters scientific discussions, as every scientist knows. Darwin is not a scientist but a quack who uses poetry to justify his "scientific discoveries."

> We have every reason to believe that many of these doubtful and closely allied forms have permanently retained their characters for a long time; for as long, as for as we know, as have good and true species. (Darwin, *Origin*, 61)

When someone says, "we have every reason to believe . . ." and gives no explanation of why we should believe, a more appropriate response is, "we have every reason not to believe." There are no historic records that would prove that certain forms have retained their characters for a long time, so Darwin's statement is a conjecture at most, not a fact. As everyone knows, true and good science is based on facts, not on conjectures.

> Many of the cases of strongly-marked varieties or doubtful species well deserve consideration; for several interesting lines of argument, from geographical distribution, analogical variation, hybridism, etc, have been brought to bear in the attempt to determine their rank; but space does not here permit me to discuss them. (Darwin, *Origin*, 64)

In the time of Darwin, as today, several theories were put forward to explain strongly marked varieties. Determining factors such as geographical distribution, hybridism, etc., vary from one theory to another, and none of these theories has been universally accepted.

When there are several competing theories without a clear front-runner, there is a clear indication that the science in question, in this case biology, is so poorly developed that it is unsuitable for any kind of deduction and practical application.

> These facts [the plants of 12 countries and the insects of 2 districts] are of plain signification on the view that species are only strongly-marked and permanent varieties; for wherever many species of the same genus have been formed, or where, if we may use the expression, the manufactory of species has been active, we ought generally to find the manufactory still in action, more especially as we have every reason to believe the process of manufacturing new species to be a slow one. (Darwin, *Origin*, 70)

Terms like *slow process* or *fast process* are irrelevant to scientific development; there are cosmological processes that last for millions of years and it's a matter of taste whether to call them rapid or sluggish. However, every process that a scientific theory describes is required to meet the criterion of observability. Non-observable processes belong to the realm of metaphysics and are outside the realm of science.

Processes that Darwin refers to as "manufactory" of species are non-observable; so far no one has been able to see the formation of species. Calling it a "slow process" does not change a thing; they still have to meet the criterion of observability, otherwise the evolutionary theory is pure metaphysical crap and cannot be called a scientific theory.

> No naturalist pretends that all the species of a genus are equally distant from each other; they may generally be divided into sub-genera, or sections, or lesser groups. As Fries has well remarked, little groups of species are generally clustered like satellites around other species. And what are varieties but groups of forms, unequally related to each other, and clustered round certain forms—that is, round their parent-species. (Darwin, *Origin*, 71)

Are clusters of parent-species smaller or larger? How is the distance between clusters defined? What is the radius of a cluster? These are not idle questions; they are always asked when artificial intelligence specialists develop their theories. Biologists do not have algorithms of their own. For them these vague questions with even more vague answers are scientific descriptions—biologists do not have a clue about how cluster theories work. As a result, their theories are nothing more than a stream of pseudo-scientific jargon.

## Unnatural Selection

> I should premise that I use this term [struggle] in a large and metaphorical sense including dependence of one being on another, and including (which is more important) not only the life of the individual, but success in living progeny. (Darwin, *Origin*, 74)

Anything could be used in a metaphorical sense; however, science deals not with metaphors but with precise definitions—this is one of the major differences between science and poetry.

> Every being, which during its natural lifetime produces several eggs or seeds, must suffer destruction during some period of its life; and during some season or occasional year; otherwise, on the principle of geometrical increase, its numbers would quickly become so inordinately great that no country could support the product . . .
>
> There is no exception to the rule that every organic being naturally increases at so high a rate, that, if not destroyed, the earth would soon be covered by the progeny of a single pair. (Darwin, *Origin*, 75)

This, Darwin's "masterpiece," demonstrates everything that is wrong with biological science; its lack of clarity is astonishing. What does the word *destruction* mean in this context? If it means death due to natural causes, then there is no struggle for survival involved. Could it mean that some species are attacked and destroyed by other species? But many species, sharks for example, do not have natural enemies, so the concept of destruction is not applicable to them. Perhaps it means that some individuals die because of starvation? But then the destruction is too narrowly defined to draw a conclusion.

Darwin's math is also pathetic—geometrical increase can occur only when a species is immortal. Darwin is completely unfamiliar with mathematical modeling; otherwise, he would have known that under certain conditions it is possible that not a single species would become extinct due to overpopulation, the condition being that the average life span of the species is shorter than the time needed to reach reproductive age, as in several types of fish and bacteria. Apparently, there are exceptions to Darwin's rule.

> Our familiarity with the large domestic animals tends, I think, to mislead us: we see no great distinction falling on them, but we do not keep in mind that thousands are annually slaughtered for food, and that in a state of nature an equal number would have somehow to be disposed of. (Darwin, *Origin*, 77)

Yes, we slaughter poor cows, but in India the cow is a sacred animal (the funny American expression "Holy Cow" is, in fact, a reference to Hinduism). The killing of a cow is a terrible crime against the god Krishna. Then again, the Indian state of nature is different from the British state of nature—it does not destroy cows.

> The causes which check the natural tendency of each species to increase are most obscure. Look at the most vigorous species; by as much as it swarms in numbers, by so much will it tend to increase still further. We know not what exactly the checks are even in a single instance. (Darwin, *Origin*, 78)

Every scientific theory begins by stating its objectives. In the case of the evolutionary theory, one of the objectives would be the determination of causes that keep the tendency to increase in check. As Darwin candidly admits, the causes are obscure. But this implies that the theory never got off the ground; it remains a hypothesis at most. Darwin's admission is equivalent to saying that no progress has been made so far that would establish the evolutionary theory as a full-grown scientific theory.

> That climate acts in main part indirectly by favoring other species; we clearly see in the prodigious number of plants which in our gardens can perfectly well endure our climate, but which never become naturalized, for they cannot compete with our native plants nor resist destruction by our native animals. (Darwin, *Origin*, 79)

Somehow the potato got naturalized throughout Europe and North America; perhaps native animals do not find it tasty. The same is true for many other naturalized species. But Darwin excludes everything that does not fit his climatic hypothesis—such exclusions are the hallmarks of the evolutionary theory.

> When a species, owing to highly favorable circumstances, increases inordinately in numbers in a small tract, epidemics—at least, this seems generally to occur with our game animals—often ensue; and here we have a limiting check independent of the struggle for life. But even some of these so-called epidemics appear to be due to parasitic worms, which have from some cause, possibly in part through facility of diffusion amongst the crowded animals, been disproportionately favored: and here comes in a sort of struggle between the parasite and its prey. (Darwin, *Origin*, 79)

The crowding of domestic animals could cause the spread of parasites, but this scenario is unlikely to occur in the domain of wild animals because of the abundance of open spaces. Epidemics in wild habitats are caused by viruses and bacteria, as the spread of the bird flu shows. When an epidemic is caused by microorganisms, the number of species is irrelevant, as all epidemiologists know. Therefore, there are natural mechanisms other than the struggle for life that determine the fate of a species.

> The dependency of one organic being on another, as of a parasite on its prey, lies generally between beings remote in the scale of nature. This is likewise sometimes the case with those which may strictly be said to struggle with each other for existence, as in the case of locusts and grass-feeding quadrupeds. But the struggle will almost inevitably be most severe between the individuals of the same species, for they frequent the same districts, require the same food, and are exposed to the same dangers. (Darwin, *Origin*, 80)

If Darwin is correct, more antelopes kill one another than fall victim to lions. This is quite a stretch of the imagination! Perhaps there are serial killers among the herd who murder the offspring! Do biologists realize how ridiculous Darwin's statement is? There is another problem with this paragraph: it contradicts the preceding paragraph. Maybe it is not a parasite at all that kills animals living in overcrowded conditions.

> In the case of every species, many different checks, acting
> at different periods of life, and during different seasons or
> years, probably come into play; some one check or some few
> checks being generally the most potent, but all will concur
> in determining the average number or even the existence of
> the species. (Darwin, *Origin*, 83)

From a theoretical standpoint, in order to determine the average
number of a species one must know the most important constraints,
otherwise there is no way of knowing if the proposed average number
is correct. Of course, it is possible to determine the average number
by direct counting, but in the absence of a theory this number would
be a pure statistic and would represent nothing. Darwin suggests that
there are checks, or constraints, but provides no criteria for selecting
the most important ones. His statement is detached from reality in
the sense that it cannot be verified.

> On the confines of its geographical range, a change of consti-
> tution with respect to climate would clearly be an advantage
> to our plant; but we have reason to believe that only a few
> plants or animals range so far, that they are destroyed exclu-
> sively by the rigour of the climate. (Darwin, *Origin*, 85)

In this unintelligible text Darwin is trying to say that only a few
plant and animal species are not destroyed by the climate outside their
native territory, while the others undergo destruction.

What makes an indestructible species different from the rest of
the field? There is no explanation; more importantly, there is no way
of telling how many exceptions there are other than saying that exceptions are few. From a logical standpoint, a rule with exceptions is not a
rule but a factual statement. It is permissible to use factual statements,
or statistics, as far as descriptions of natural events go, but a theory
without rules is completely useless because it does not allow for any
kind of prediction. Somehow biologists have forgotten that the aim
of a scientific theory is not only to put events into categories but to
predict them as well.

## Survival of the Non-fittest

> Can it, then, be thought improbable, seeing that varia-
> tions useful to man have undoubtedly occurred, that other

variations useful in some way to each being in the great and complex battle of life, should occur in the course of many successive generations. If such do occur, can we doubt (remembering that many more individuals are born than can possibly survive) that individuals having any advantage, however slight, over others, would have the best chance of surviving and of procreating their kind? On the other hand, we may feel sure that any variation in the least degree injurious would be rigidly destroyed. This preservation of favourable individual differences and variations, and the destruction of those which are injurious, I call Natural Selection, or the Survival of the Fittest. (Darwin, *Origin*, 87)

What happens if an eagle breaks one of his claws? The injury would be so slight that it would be unlikely to put him at a terrible disadvantage. Most likely, his life would go on as if nothing had happened. Now, compare two eagles, one with a broken claw and living in an area with an abundance of food, and the other one with his claws intact but living in an area where food is scarce. Which one of them has a better chance of survival?

One might argue that the principle of natural selection applies to the members of a species living in the same area, but this is not what Darwin proposes. According to him, this principle is universal; otherwise, there would be no guarantee that the fittest individuals survive and procreate.

Variations neither useful nor injurious would not be affected by natural selection, and would be left either a fluctuating element, as perhaps we see in certain polymorphic species, or would ultimately become fixed, owing to the nature of the organism and the nature of the conditions. (Darwin, *Origin*, 88)

As usual, Darwin offers no criteria that would allow one to determine whether a variation is useful or not; neither does he provide any proof that useless variations exist. Perhaps there are such things as useless variations, but we will never know for sure.

Let's assume for the sake of argument that non-harmful variations exist. Then there must be another natural mechanism different from natural selection that causes them. Perhaps this mechanism is much more important than natural selection. If this is true, then the whole evolutionary theory goes through the window. Well, the theory is already gone; this is just another reason why it should not return.

> We shall best understand the probable course of natural selection by taking the case of a country undergoing some slight physical change, for instance, of climate. The proportional numbers of its inhabitants will almost immediately undergo a change, and some species will probably become extinct. (Darwin, *Origin*, 88)

At first glance it looks as though Darwin is proposing an experiment designed to prove that the principle of natural selection exists. However, it takes several thousand years for visible climatic changes to develop, so this is an unrealistic proposition. It is impossible to design an experiment that would prove that natural selection is not a myth because it is almost always the case that a decrease in the number of a species occurs not because of natural causes, but because of environmental pollution including oil spills, indiscriminate use of pesticides, etc. When an oil spill hits the shore, all birds in the area die, not just the ones that are less fit to survive.

Those principles that defy experimental verification, the principle of natural selection being one of them, are prohibited from forming the basis of a scientific theory.

> She [Nature] can act on every internal organ, on every shade of constitutional difference, on the whole machinery of life. Man selects only for his own good: Nature only for that of the being which she tends. Every selected character is fully exercised by her; as is implied by the fact of their creation. (Darwin, *Origin*, 90)

Does nature make plans or shoot at random, so to speak? If there is planning involved, it leads to the topic of there being a creator, and this is not what the evolutionists are willing to consider. If she moves at random, then there is no selection because selection is not a random act. Could it be something between planned events and random ones? According to the norms of the English language, or any other language, there is no such middle ground.

> Nor ought we to think that the occasional destruction of an animal of any particular colour would produce little effect: we should remember how essential it is in a flock of white

sheep to destroy a lamb with the faintest trace of black.
(Darwin, *Origin*, 91)

Playing off of the well-known English song, "Baa, Baa, Black
Sheep," what if the garment industry wants to shift from white sweat-
ers to black ones? Then the black sheep is saved! This example proves
nothing. Darwin simply runs out of examples that he believes that
can support his theory.

> Natural selection will modify the structure of the young in
> relation to the parent, and of the parent in relation to the
> young. (Darwin, *Origin*, 92)

There is no such thing as identical beings, so the young is already
modified in relation to the parent. Darwin makes no suggestion of
how to distinguish changes in structure due to natural selection from
changes due to other causes, so his statement is unverifiable.

> This leads me to say a few words on what I have called
> Sexual Selection. This form of selection depends, not on a
> struggle for existence in relation to other organic beings or
> to external conditions, but on a struggle between the indi-
> viduals of one sex, generally the males, for the possession of
> the other sex.
> Generally, the most vigorous males, those which are best
> fitted for their places in nature, will leave most progeny.
> (Darwin, *Origin*, 94)

The strongest, most vigorous males spread their seed around and
the whole herd benefits. Right? Wrong! Along with physical strength
they may be spreading around such imperfections as color blindness,
albinism, high blood pressure, etc., so the herd is better off without
them. There is no guarantee that physical strength will bring about
an improvement in general, overall health. The principle of natural
selection makes sense only if the male struggle improves several vital
characteristics, which is clearly not the case.

> Let us take the case of a wolf, which preys on various ani-
> mals, securing some by craft, some by strength, and some
> by fleetness; and let us suppose that the fleetest prey, a deer
> for instance, had from any change in the country increased
> in numbers, or that other prey had decreased in numbers,
> during that season of the year when the wolf was hardest

pressed for food. Under such circumstances the swiftest and slimmest wolves would have the best chance of surviving and so be preserved or selected—provided always that they retained strength to master their pray at this or some other period of the year, when they were compelled to prey on other animals. (Darwin, *Origin*, 95)

Are all members of a species equally smart? According to the latest studies, there are considerable differences. Some individuals were even called "Einstein of the animal world." The smartest wolves are not necessarily the swiftest, but their intellectual superiority is far more important than swiftness. Unfortunately, their superiority is not transferred to the next generation, so the herd is not getting smarter.

There is another aspect that eluded Darwin's attention—wolves often hunt in packs, and the pack shares the food, so no one, except for Darwin's theory, suffers. In any case, because speed measurements of multiple wolf generations do not exist, there is no observable data to back up Darwin's propositions. But without experimental data a theory is dead upon arrival.

In such cases, if the varying individual did not actually transmit to its offspring its newly-acquired character, it would undoubtedly transmit to them, as long as the existing conditions remained the same, a still stronger tendency to vary in the same manner. (Darwin, *Origin*, 97)

In general, scientific observations are done to determine whether or not there are changes in the object of experimentation due to changes in conditions; a mysterious quality that Darwin calls "tendency" is undetectable in principle. As practically every scientist knows, changes that are impossible to detect in principle do not belong in the world of science. But, then, biologists are not true scientists; their pseudo-science encourages the inclusion of non-observable events in their silly theories.

[I]f any one species does not become modified and improved in a corresponding degree with its competitors, it will be exterminated. (Darwin, *Origin*, 104)

Certainly, some species, the dinosaurs for example, became extinct because they could not adapt to new conditions. But is this rule always applicable? It appears that certain species, the mammoth for example, became extinct when there were no changes in external conditions (the evolutionists believe that other causes of variation are random mutations, which are internal conditions; then there is also inbreeding, which can possibly cause genetic deterioration). To answer this question, one must travel back in time to ascertain what exactly were the conditions that existed prior to the extinction of a species to see if the members of that species could have survived—this is called a direct experiment. But time travel is impossible, so there is no way of saying with complete certainty what had happened to the species under observation.

> Isolation, also, is an important element in the modification of species through natural selection. In a confined or isolated area, if not very large, the organic and inorganic conditions of life will generally be almost uniform, so that natural selection will tend to modify all the varying individuals of the same species in the same manner. (Darwin, *Origin*, 106)

There are three objections to this assessment:

1) It is extremely difficult to find such isolated areas; even species living on an island tend to migrate to other islands. There may not be enough data to draw a definitive conclusion based on parsimonious statistical data.

2) People living in close communities tend to develop certain genetic disorders due to inbreeding; as the Amish example shows, this is also true for the animals. In case of animals, it is impossible to tell whether the changes in constitution are caused by natural selection or multiple inbreeding.

3) Darwin did not bother to define with required precision the term *uniform*. Without proper definition, it is not clear what kind of conditions and variations to look for.

> Although isolation is of great importance in the production of new species, on the whole I am inclined to believe that largeness of area is still more important, especially for the production of species which shall prove capable of enduring for a long period, and of spreading widely. Throughout a great and open area, not only will there be a better chance of favourable variations, arising from the large number of individuals of the same species there supported, but the

> conditions of life are much more complex from the large
> number of already existing species; and if some of these
> many species become modifies and improved, others will
> have to be improved in a corresponding degree, or they will
> be exterminated. (Darwin, *Origin*, 107)

If you roll a die, what is the chance that you will get a particular number? It is 1/6. If you roll two dice, what is the chance that they will both show the same number? By the law of independent probabilities, it is (1/6) x (1/6), which amounts to 1/36. In general, the probability that N dice land on the same number is (1/6)N.

This probability goes down fast, for where N=10 it is practically zero. The same is true for the members of a species—the chance that enough of them will get the same characteristic to spread to the entire population is practically zero; it doesn't matter whether or not the area is isolated. Darwin knew nothing about the basics of mathematical statistics. Because of a lack of knowledge he came up with all kinds of idiotic hypotheses.

> Rarity, as geology tells us, is the precursor to extinction.
> We can see that any form which is represented by few in-
> dividuals will run a good chance of utter extinction, during
> great fluctuations in the nature of the seasons, or from a
> temporary increase in the number of its enemies. (Darwin,
> *Origin*, 110)

Dinosaurs were the most populous species and yet they completely disappeared due to climatic changes, while a less populous group, the mammals, survived, as geology tells us. But Darwin is not listening. In modern times a lot of species, both big and small in number, became either extinct or are on the brink of extinction not because of some climatic changes but because of man's activities. Darwin's assessment is completely wrong; he cannot defend his favorite child, natural selection. Dead is the child!

> Again, we may suppose that at an early period of history, the
> men of one nation or district required swifter horses, whilst
> those of another required stronger and bulkier horses. The
> early differences would be very slight; but, in the course of
> time, from the continued selection of swifter horses in the
> one case, and of stronger ones in the other, the differences

would become greater, and would be noted as forming two
sub-breeds. (Darwin, *Origin*, 111)

Why has no one tried to breed a horse that is both swift and
strong at the same time? Physiologically, nothing indicates that it cannot be done. Imagine how rich a person breeding such horses could
become! If there is an obstacle on the road to riches, that would be
the evolutionary theory.

All horse-breeders know that there is only one fastest-running
breed of horses, sometimes called race horses; the others do not even
come close. There have been numerous attempts to bring other horse
species up to race-horse speeds, but they have all failed, although
Darwin predicted otherwise.

> Consequently, each new variety of species, during the
> progress of its formation, will generally press hardest on its
> nearest kindred, and tend to exterminate them. We see the
> same process of extermination amongst our domesticated
> productions, throughout the selection of improved forms
> by man. . . .
> In Yorkshire, it is historically known that the ancient
> black cattle were displaced by the long-horns, and that these
> "were swept away by the short-horns (I quote the words of
> an agricultural writer) as if by some murderous pestilence."
> (Darwin, *Origin*, 110)

There are certain species of monkeys living in close proximity
to one another. If Darwin is correct, they should have exterminated
at least some of their cousins, and yet it has never happened. Perhaps
their process of formation has already stopped, as modern-day evolutionists believe. But then why has the formation process hit a snag?
There is no answer to this question because there is no formation
process!

In Yorkshire, not enough time had passed to produce new cattle
species by natural selection; according to the evolutionists that takes
several thousands years. The Yorkshire cattle were replaced by imported specimens, as historic records show.

> Nevertheless, according to my view, varieties are species in
> the process of formation, or are, as I have called them, in-
> cipient species. (Darwin, *Origin*, 111)

Is it possible to determine whether the formation of a new species is taking place? Yes, it is, and the procedure is relatively simple: all one has to do is to measure a number of physical characteristics in several generations, find the average value for each characteristic, and then determine whether the differences between the average values are statistically significant. Nothing of this sort has been done so far; otherwise, there would have been plenty of reports confirming that the process of formation is not dead cold. But because of the absence of observations, no conclusion can be drawn. Darwin is kind of a weird scientist—he draws conclusions in cases where no one else dares to.

> Here, then, we see in man's productions the action of what may be called the principle of divergence, causing differences, at first barely appreciable, steadily to increase; and the breeds to diverge in character, both from each other and from their common parent. (Darwin, *Origin*, 112)

Is it possible to say with absolute certainty which are the breeds and which are the parents? After all, they all look alike and there is no historic record to tell which breed is the oldest. Assuming that there is a parent, this is an impossible task.

Evolutionists say that in many cases the parent became extinct. What is the difference, then, between an extinct species and a species that exists only in the evolutionists' minds? None whatsoever! The whole paragraph is a meaningless conflation of the words *breeds* and *parent*.

*Meaningless* is a word equally applicable to the evolutionary theory with its unproven assertions and vague definitions.

# 22 : Death Due to Diversification

> The truth of the principle that the greatest amount of life can be supported by great diversification of structure, is seen under many natural circumstances. (Darwin, *Origin*, 113)

What if, as a result of diversification, an extremely aggressive species that kills all other species in its habitat were to appear? (This is the story of several imported species.) Then diversification equals death! Darwin did not consider this scenario; otherwise he may have known that it leads to the existence of a much smaller number of species than currently exists.

> Take the case of a carnivorous quadruped, of which the number that can be supported in any country has long ago arrived at its full average. If its natural power of increase be allowed to act, it can succeed in increasing (the country not undergoing any change in conditions) only by its varying descendants seizing on places at present occupied by other animals: some of them, for instance, being enabled to feed on new kinds of prey, either dead or alive; some inhabiting new stations, climbing trees, frequenting water, and some perhaps becoming less carnivorous. (Darwin, *Origin*, 112)

A few pages earlier Darwin insists that variations in species are caused by changes in conditions. Now he suggests that, due to a mysterious "natural power of increase," it is possible to have variations without changes in conditions. This is just another example of the type of contradictory statements that the evolutionary theory is full of.

It seems strange that any physical, chemical, or biological process would continue to vary under static conditions; actually, this is unheard of. Darwin proposes something that any other scientist would call crazy talk—one of the basic premises of any science is the assumption of permanency of the results of an experiment under unchanged conditions. (In quantum mechanics, the results may vary according to probabilistic laws, but the wave-function is always the same, which indicates that the cause-and-effect relationship is still valid. Darwin, on the other hand, denies that this relationship exists.)

Everything is possible in Shangri-La: wolves climbing trees, tigers swimming with fishes, lions eating tomatoes. Darwin is such a great storyteller! In reality even the evolutionists with their overactive imaginations would find it hard to believe in Darwin's picture of the not-so-distant past.

> By considering the nature of the plants or animals which have in any country struggled successfully with the indigenes and have there become naturalised, we may gain some crude idea in what manner some of the natives would have to be modified, in order to gain an advantage over their compatriots . . . (Darwin, *Origin*, 114)

In the beginning of his book, Darwin says that naturalization is impossible (see the previous chapter); now he blatantly contradicts himself. What's going on? Apparently, Darwin has a very short memory. Because of that his theory falls short of making sense.

> The accompanying diagram will aid in understanding this rather perplexing subject. Let A to L represent the species of a genus large in its own country; these species are supposed to resemble each other in unequal degrees, as is so generally the case in nature, and as is represented in the diagram by the letters standing at unequal distances. (Darwin, *Origin*, 115)

In scientific papers, diagrams are drawn to explicitly or implicitly represent numeric values; phrases like "unequal degree" are never used because of their vagueness and the impossibility of representing them on a graph. Darwin wasted several pages of his book trying to explain his perplexing diagram that represents nothing. This is another pathetic attempt of his to put pseudo-scientific ideas into scientific form.

> The intervals between the horizontal lines in the diagram may represent each a thousand or more generations. After a thousand generations, species (A) is supposed to have produced two fairly well-marked varieties. . . . Moreover, these two varieties, being only slightly modified forms, will tend to inherit those advantages which made their parent

(A) more numerous than most of the other inhabitants of the same country. . . . (Darwin, *Origin*, 116)

A thousand or more? How much more? Ten million perhaps? In this case, there is no historic data to support the theory.

Are "well-marked varieties" the same as "slightly modified forms"? According to the rules of any language, they are not; Darwin produced two contradictory statements in a short paragraph. If both varieties inherited the advantages that made their parents very numerous, then why did they split into two groups? It seems logical to continue the parental line without a split.

> Nor do I suppose that the most divergent varieties are invariably preserved: a medium form may often long endure, and may or may not produce more than one modified descendant; for natural selection will always acts according to the nature of the places which are either unoccupied or not perfectly occupied by other beings; and this will depend on infinitely complex relations. (Darwin, *Origin*, 117)

First, Darwin says that diversification is the key to survival; now he insists that the most divergent varieties may not be preserved—this is an infinitely confusing statement. According to Darwin, they may fail to survive because of natural selection, and at the same time he believes that natural selection produces a variety of species by preserving the most advantageous characteristics. It almost seems as though the phrase "cause-and-effect relation" means nothing to Darwin.

> The modified offspring from the later and more highly improved branches in the lines of descent, will, it is probable, often take the place of, and so destroy, the earlier and less improved branches . . . (Darwin, *Origin*, 117)

Exactly how is this destruction carried out? Theoretically, there are two ways of achieving it:

1) *The less advanced offspring is pushed out of its natural habitat into the areas where the other species kill it.*

Bad what if the "bad" offspring are carnivorous and all species in the new habitat are slow-moving herbivores? Then destruction is impossible, unless Darwin is waiting there with a hunting rifle!

2) *The improved offspring kills its cousins.*

This would imply that the improved offspring is physically stronger. However, physical strength may not be the most desirable characteristic. In fact, it may be the least desirable characteristic because it leads to more severe injuries when the males of a more advanced branch fight amongst themselves.

Darwin jumped to unwarranted conclusions because these possibilities never crossed his mind.

> For it should be remembered that the competition will generally be most severe between those forms which are most nearly related to each other in habits, constitution and structure. (Darwin, *Origin*, 119)

Is it possible for two or more related species to move away from each other and avoid competition? Of course it is—there are plenty of migratory species that do just that. But this implies that the principle of natural selection does not apply to them.

> I see no reason to limit the process of modification, as now explained, to the formation of genera alone. . . . We shall also have two very distinct genera descended from (I), differing widely from the descendents of (A). These two groups of genera will thus form two distinct families, or orders, according to the amount of divergent modification supposed to be represented in the diagram. (Darwin, *Origin*, 121)

If they differ widely from each other, how do we know that they have the same parent? There is no guarantee that species (A) and (I) are completely unrelated to each other, so the diagram is practically useless. This "science" is inexact to such a degree that no amount of semantics can hide the fact that species having nothing in common are likely to be classified under the same parent. Wouldn't it be more logical to suggest, because of the impossibility of finding the parent of two or more species, that there is no parent?

> Looking to the future, we can predict that the groups of organic beings which are now large and triumphant, and which are least broken up, that is, which have as yet suffered least extinction, will, for a long period, continue to increase. (Darwin, *Origin*, 122)

Initially, Darwin says that there are checks, or constraints, that prevent unlimited growth from happening (see the previous chapter); now he predicts substantial growth. Which one is true? This is a win-win situation for him—no matter what the observational data is, the evolutionary theory wins! What better way to build a theory that is always correct than by making it self-contradictory!

> Looking still more remotely to the future, we may predict that, owing to the continued and steady increase of the larger groups, a multitude of smaller groups will become utterly extinct. (Darwin, *Origin*, 122)

Yes, they become extinct, but not because of an increase of larger groups; rather, it is predominantly because of environmental pollution and poaching. This is a sad fact of our times, but it also shows that Darwin's theory failed to make correct predictions.

> Among the vertebrata the degree of intellect and an approach in structure to man clearly come into play. (Darwin, *Origin*, 122)

So, according to Darwin, natural selection not only improves the organism's structure but also increases its intellectual capabilities. Darwin also believed that natural selection never ceases its work, so the conclusion is clear—judging by the ape's example, the other vertebrata, such as lions, kangaroos, carps, etc., should have intelligence equal to that of man (by now Superman should be fighting Aquaman). Not in our lifetime. Modern-day evolutionists tried to bypass this difficulty by saying that natural selection works only when conditions are ripe. It better start working now because environmental pollution is killing dozens of species every year. Environmentalists may be praying for natural selection to wake up from its slumber, but evolutionist theories are such a sleeping pill!

> On the other hand, we can see . . . that it is quite possible for natural selection gradually to fit a being to a situation in which several organs would be superfluous or useless: in such cases there would be retrogression in the scale of organization. (Darwin, *Origin*, 123)

In chapter 1 of his book Darwin expressed admiration for nature's striving for perfection (see the previous chapter of this book); now he suggests that nature is capable of horrible imperfection and retrogression. But in the context of Darwin's theories, retrogression ultimately implies extinction because the species who suffer it are no longer the fittest. Therefore, we come to a paradoxical conclusion: natural selection causes extinction. Clearly, Darwin is at war with himself.

> The direct action of changed conditions leads to definite or indefinite results. In the latter case the organization seems to become plastic, and we have much fluctuating variability. In the former case the nature of the organism is such that it yields readily, when subjected to certain conditions, and all, or nearly all the individuals become modified in the same way. (Darwin, *Origin*, 131)

There are two kinds of events that scientists encounter: deterministic and probabilistic. Both types produce definite results because they can be recorded. Therefore, the fluctuations Darwin refers to are definite events. It follows that no matter what the conditions are, all individuals become modified in the same way. But this is not what experiments show.

What are the indefinite results anyway? The word *indefinite* is very seldom used in science to designate those events that cannot be observed. Then again, as almost every scientist knows, such events do not exist anyway.

> It is very difficult to decide how far changed conditions, such as of climate, food, &c., have acted in a definite manner. There is reason to believe that in the course of time the effects have been greater than can be proved by clear evidence. (Darwin, *Origin*, 131)

In other words, the evolutionary theory proves that the experimental data is wrong! How is that possible? Certain Greek philosophers, including Plato, were known for their disregard of experimental data when it contradicted their logical deduction, which they believed to be the supreme judge, but those days of crazy science are long gone. Darwin tried to bring them back, but even the evolutionists must admit that he went out on a limb with this one.

> When a variation is of the slightest use to any being, we cannot tell how much to attribute to the accumulative action

of natural selection, and how much to the definite action of
the conditions of life. (Darwin, *Origin*, 132)

An accumulative effect is the sum of effects that occur over the
course of generations, each of those individual effects being so small
that the cause is unknown. This implies that it is impossible to tell
whether the cumulative effect is due to natural selection or to conditions of life, or both, or neither. The whole theory is invalid because
it defies verification.

> I think there can be no doubt that use in our domestic
> animals has strengthen and enlarged certain parts, and
> disuse diminished them; and that such modifications
> are inherited under free nature, we have no standard
> of comparison, by which to judge of the effects of long-
> continued use or disuse, for we know not the parent-forms;
> but many animals possess structures which can be best
> explained by the effects of disuse. As Professor Owen has
> remarked, there is no greater anomaly in nature than a bird
> that cannot fly; yet there are several in this state. (Darwin,
> *Origin*, 133)

I'll start with Darwin's startling admission: "we know not the
parent-forms." From a logical standpoint, this remark is equivalent to
saying "I do not know if the evolutionary theory is correct because I
cannot prove any of its assertions." Darwin, then, seems to contradict
himself, as he does on numerous occasions, by implying that in the
case of domestic animals the parents are known. Why is it so easy to
trace the descent line in the case of domestic animals and impossible
in the case of wild ones? No one knows!

Now, to the birds with underdeveloped wings—is this the result
of man's selection? Let me ask the following question: what is the
most important purpose of wings? Darwin and Owen believed that
it was flight, but this is incorrect. The most important usage of wings
occurs during the incubation period (Owen tried to warm up cuckoo
eggs with his hands, but it was not nearly as effective as the cuckoo's
wings). Wings are also needed for proper blood circulation in the
bird's body and for the production of certain ferments, so an inability
to fly does not prove that domestic birds underutilize their wings.

> The ostrich indeed inhabits continents, and is exposed to danger from which it cannot escape by flight, but it can defend itself by kicking its enemies, as efficiently as many quadrupeds. We may believe that that the progenitor of the ostrich genus had habits like those of the bustard, and that, as the size and weight of its body were increased during successive generations, its legs were used more, and the wings less, until they became incapable of flight. (Darwin, *Origin*, 138)

Presumably, the ostrich's progenitor hated being victimized by birds of prey. But the progenitor was probably not the only bird living in the area. It would seem natural for the other bird species to follow the ostrich's example and start kicking and screaming, but they didn't. The ostrich is not the smartest bird—it hides its head in the sand—and yet it managed to outsmart its fellow species. How smart were they? We are talking about bird's brains!

Why wouldn't the progenitor learn to kick while keeping its wings? Just imagine an ostrich kicking a hawk in the air—one kick and the evolutionary theory is dead!

> Kirby has remarked (and I have observed the same fact) that the anterior tarsi, or feet, of many male dung-feeding beetles are often broken off . . . In the Onites apelles the tarsi are so habitually lost, that the insect has been described as not having them. In some other genera they are present, but in a rudimentary condition. In the Ateuchus or sacred beetle of the Egyptians, they are totally deficient . . . Hence it will perhaps be safest to look at the entire absence of the anterior tarsi in Ateuchus, and their rudimentary condition in some other genera, not as cases of inherited mutilations, but as due to the effects of long-continued disuse . . . (Darwin, *Origin*, 134)

A glass and a cup have one major difference: a cap has a handle while a glass does not. Is it correct to say that some cups lost their handles due to disuse and turned into glasses? Darwin uses the same logic to infer that certain types of beetles lost their anterior tarsi, but, as the evidence in the case of Onites apelles suggests, they were likely not present in the first place. Why did such bad things happen to the male beetles only, while the females managed to keep their feet intact even living in the same conditions? There is no answer to that.

> It is well known that several animals, belonging to the most different classes, which inhabit the caves of Carniola and of Kentucky, are blind. In some of the crabs the foot-stalk for the eye remains, though the eye is gone;—the stand for the telescope is there, though the telescope with its glasses has been lost. As it is difficult to imagine that eyes, though useless, could be in any way injurious to animals living in darkness, their loss may be attributed to disuse. (Darwin, *Origin*, 135)

It is not only difficult but impossible to imagine how eyes could be injurious to anyone—they are, after all, the most important organs of perception. How could disuse of the eyes possibly happen? There is only one way: by keeping them constantly closed. Says the predator: "Thank you for keeping your eyes closed, silly crab!"

> As it is extremely common for distinct species belonging to the same genus to inhabit hot and cold countries, if it be true that all the species of the same genus are descended from a single parent-form, acclimatisation must be readily effected during a long course of descent. (Darwin, *Origin*, 137)

There are several mammal species of the same genus living on different continents. How did the migration occur? Usually, evolutionists answer this question by saying that in the very distant past the continents were positioned very close to one another. Together they basically constituted one continent, then drifted apart. But that happened before the extinction of dinosaurs, and by all geological accounts there were no signs of mammals at that time (by the way, a transition form the dinosaur to the mammal is impossible for physiological reasons: these species exhibit fundamental differences in cardiovascular systems). Darwin and his contemporaries knew very little about dinosaurs due to the lack of paleontological data. They assumed that mammals and birds always existed and their assumption led to an error of continental proportions.

> We have reason to believe that species in a state of nature are closely limited in their ranges by the competition of other

organic beings quite as much as, or more than, by adaptation to particular climates. (Darwin, *Origin*, 137)

This paragraph, which states that species are limited in their
range of travel and, therefore, have adapted to a particular climate,
is in direct contradiction with the preceding paragraph, which states
that species can migrate across vast geographic areas, thus allowing a
single parent-form to give rise to a whole genus. Darwin makes these
contradictory statements on the same page of his book! What was
Darwin's state of mind when he was writing his pathetic book?

> *Correlated variation.* I mean by this expression that the
> whole organisation is so tied together during its growth and
> development, that when slight variations in any one part
> occur, and are accumulated through natural selection, other
> parts become modified . . .
>
> We may often falsely attribute to correlated variation
> structures which are common to whole groups of species;
> and which in truth are simply due to inheritance; for an
> ancient progenitor may have acquired through natural
> selection some one modification in structure, and, after
> thousands of generations, some other and independent
> modification; and these two modifications, having been
> transmitted to a whole group of descendants with diverse
> habits, would naturally be thought to be in some necessary
> manner correlated. (Darwin, *Origin*, 141)

In the beginning Darwin defines correlated variations as correlated changes in structure being caused by natural selection, then
remarks that natural selection also produces independent variations.
Is it possible for a single cause to produce two diametrically opposed
outcomes? This is a blatant violation of all rules of logic; it automati-
cally invalidates all of Darwin's logical deductions. Even if we assume
that natural selection produces only one type of change, say, correlated variations, we are faced with the problem of discerning correlated
changes from independent ones. In mathematical statistics there is
a procedure that enables one to determine whether the variables in
question are independent or not, and it is based on known frequencies of possible outcomes. But the evolutionary theory provides no
means to determine those frequencies, thereby making it impossible
to state whether changes in structure are correlated or not.

The elder Geoffrey and Goethe propounded, at about the same time, their law of compensation or balancement of growth; or, as Goethe expressed it, "in order to spend on one side, nature is forced to economise on the other side." I think this holds true to a certain extent with our domestic productions: if nourishment flows to one part or organ in excess, it rarely flows, at least in excess, to another part; thus it is difficult to get a cow to give much milk and to fatten readily. (Darwin, *Origin*, 142)

It seems strange that a scientist would even mention a law so ridiculous that not a single textbook would ever present it. Goethe was a poet and he probably believed in the law of poetic justice, but Darwin claims to be a scientist, or at least pretends to be. But even the pseudo-scientists do not consider this law seriously. Darwin, on the other hand, based his evolutionary theory on this law, which clearly implies that natural evolution eliminates or severely disfigures organs that are not vital to the survival of a species. But if this were true, stallions would have lost their manes, dogs would have become tailless, etc., because these "appendages" are not essential for survival.

Rudimentary parts, as it is generally admitted, are apt to be highly variable. We shall have to recur to this subject; and I will here only add that their variability seems to result from their uselessness, and consequently from natural selection having had no power to check deviations in their structure. (Darwin, *Origin*, 144)

Let's juxtapose two statements: a) uselessness causes variability in rudimentary parts and b) uselessness causes permanence in rudimentary parts. It is impossible to tell which one of them is correct because they are not based on a theory or even a hypothesis. They both hang unsupported in the air, just like the evolutionary theory does.

What is a rudimentary part anyway? Perhaps it is something without which an organism can get by. But both humans and animals can get by without very important parts such as legs, eyes, ears, etc., so this definition cannot be right. Maybe rudimentary parts are the parts not engaged in any function. This definition seems correct, but in reality the science of physiology cannot say with absolute certainty that such parts exist. Remember the story of the appendix? For a long period of time it was thought that it does nothing, which would put it into the category of rudimentary parts. But not long ago it was discovered that the appendix functions as a kind of filter, thus rendering it anything but rudimentary.

> When a part has been developed in an extraordinary manner in any one species, compared with the other species of the same genus, we may conclude that this part has undergone an extraordinary amount of modification since the period when the several species branched off from the common progenitor of the genus. (Darwin, *Origin*, 145)

As biologists know, many species belonging to a single genus live in basically the same areas. So, why would such "extraordinary" development happen only in some of the animals and not in all of them? More importantly, why did the branching off occur? After all, they were all subjected to the same conditions.

Do seemingly similar species always belong within the same genus? The anteater and the elephant may look somewhat alike, but categorizing them within the same genus is like putting round white onions and white apples in the same genus (it is, of course, not the case that apples and onions belong within the same genus). Biologists' division into categories is so arbitrary that often species having almost nothing in common are put in one group, making it look as though extraordinary variations do indeed exist within species belonging to a single genus.

> It is notorious that specific characters are more variable than generic. To explain by a simple example what is meant: if in a large genus of plants some species had blue flowers and some had red, the colour would be only a specific character, and no one would be surprised at one of the blue species varying into red, or conversely; but if all the species had blue flowers, the colour would become a generic character, and its variation would be a more unusual circumstance. (Darwin, *Origin*, 147)

Darwin failed to make a distinction between generic and specific characteristics. As a result, his argument about variability is meaningless. This can be seen in the following: a flower may or may not undergo changes in its genetic makeup, but if a change has occurred, it affects the descendants of an individual in accordance with the laws of genetics. It seems natural to assume that if changes of color are gradual, all shades of red, green, orange, etc., in the descendants could be observed. But from an experimental standpoint, such observations are impossible to achieve because there are millions of shades, the differences between them being so small that they are practically unobservable. We have no other choice but to accept that the experiment that could confirm such gradual changes does not exist. Therefore, we have to assume that the changes were abrupt. But if a change is abrupt, then the implication is that several individuals acquired the same color simultaneously; it could not come as a result of mutation in one organism because it wouldn't propagate to the entire population (see chapter 6 of this book). But an abrupt, simultaneous change in several flowers is something unheard of.

The only remaining logically valid conclusion is that there was no variability in the plants

> On the ordinary view of each species having been independently created, why should that part of the structure, which differs from the same part in other independently-created species of the same genus, be more variable than those parts which are closely alike in the several species? I do not see that any explanation can be given. (Darwin, *Origin*, 147)

Why shouldn't they be more variable? Was it perhaps a whim on the creator's part? In any case, questions about why things are or are not built in a certain way should be expelled from the realm of science, as Niels Bohr correctly noted. Instead, scientists should be asking themselves, how are things functioning? The why question is purely metaphysical and should not be asked within science even in a rhetorical sense.

> It might further be expected that the species of the same genus would occasionally exhibit reversions to long lost characters. As, however, we do not know the common ancestors of any natural group, we cannot distinguish between reversionary and analogous characters. (Darwin, *Origin*, 152)

If species revert to long lost characteristics, then natural selection is not doing its job because undesirable characteristics are certain to appear. If they appear, this is evidence that species are not improving and the evolutionary theory is wrong.

How would we know that there are reversionary and analogous characters if we cannot distinguish them from each other? Not a single branch of natural science, except for biology, relies upon there being a distinction between characteristics that are, in fact, indistinguishable from one another. The fact that reversionary and analogous characters are indistinguishable renders the very distinction between them dubious.

> Whatever the cause may be of each slight difference between the offspring and their parents—and a cause for each must exist—we have reason to believe that it is the steady accumulation of beneficial differences which has given rise to all the more important modifications of structure in relation to the habits of each species. (Darwin, *Origin*, 157)

This is a glaring contradiction to the above quotation where the concept of reversions is introduced. If Darwin is correct about reversion, it is quite possible at least in some species that an accumulation of beneficial differences does not occur. Why did Darwin entertain the notion of reversions? Apparently he saw the lack of useful qualities in some species. This is the only possible explanation.

> But, as by this theory innumerable transitional forms must have existed, why do we not find them embedded in countless numbers in the crust of the earth? . . . I believe the answer mainly lies in the record being incomparably less perfect than is generally supposed. The crust of the earth is a vast museum; but the natural collections have been imperfectly made, and only at long intervals of time. (Darwin, *Origin*, 159)

Scientific knowledge is based on facts, not on unproven theories. As Darwin remarks, at the present time there is not enough geological data to support the evolutionary theory. Of course, it would be possible to redefine the evolutionary theory as a hypothesis and then wait for geological confirmation, but this is not what the evolutionists propose. They demand that their hypothesis be deemed a theory.

There are hundreds of thousands of species, both big and small; perhaps it would be unrealistic to expect that transitional forms of all of them could be unearthed. But skeptics are entitled to see transitional forms of at least one current species. Yet nothing of that sort has been offered thus far.

> When we see any structure highly perfected for any particular habit, as the wings of a bird for flight, we should bear in mind that animals displaying early transitional grades of the structure will seldom have survived to the present day, for they will have been supplanted by their successors, which were gradually rendered more perfect through natural selection. (Darwin, *Origin*, 163)

Without having observed transitional species, how would we know that birds descended from earth-bound creatures and not vice versa? Earlier Darwin suggested that the ostrich's progenitor was able to fly; perhaps it was birds rather than walking creatures that first inhabited the earth. By the same token, perhaps primitive organisms like amoebas, bacteria, etc., descended from birds and mammals. Anything is possible so long as we have no evidence of these alleged transitional species.

Evolutionists are certain to claim that the above proposals are wild guesses without logical foundation. Wild as they are, these guesses are no more wild than the evolutionary theory.

> As we sometimes see individuals following habits different from those proper to their species and to the other species of the same genus, we might expect that such individuals would occasionally give rise to new species, having anomalous habits, and with their structure either slightly or considerably modified from that of their type. (Darwin, *Origin*, 166)

As researchers have observed on numerous occasions, members of the same species will often approach laboratory tasks differently, some of them being more successful than others. Apparently there is an inequality in intellectual capabilities. But, as all researchers know, superior intellect is not inherited by the descendents, although other unrelated pedigrees may have their own "animal geniuses." Clearly, the fact that certain individuals behave differently than other members of the same species does not prove anything.

> He who believes that each being has been created as we now see it, must occasionally have felt surprise when he has met

with an animal having habits and structure not in agreement. What can be plainer than that the webbed feet of
ducks and geese are formed for swimming? Yet there are up-
land geese with webbed feet which rarely go near the water
. . . (Darwin, *Origin*, 167)

Apparently, Darwin is not very smart. With this remark he
delivered a fatal blow to the evolutionary theory: according to his
teachings, the goose with webbed feet who does not go near the water
should have become extinct a long time ago, having been rejected
by natural selection as an imperfect specimen. Yet such birds are still
around and torment new generations of evolutionists.

Reason tells me, that if numerous gradations form a simple
and imperfect eye to one complex and perfect can be shown
to exist . . . then the difficulty of believing that a perfect and
complex eye could be formed by natural selection, though
insuperable to our imagination, should not be considered as
subversive of the theory. (Darwin, *Origin*, 168)

Well, it is not hard to imagine a transition from an imperfect
eye to a perfect one, but what about a transition from *no eye at all*
to an eye? Presumably, the original cell did not have eyes, otherwise
its direct descendants such as viruses, amoebas, etc., would have had
eyes, too. This transition is not only hard to imagine, it's impossible
to describe, even for someone with Darwin's imagination. Actually,
Darwin tried to outline the transition, as the next paragraph in his
book shows. But he started with primitive organisms that already had
some sort of optical nerve, so this is not ground zero.

If it could be demonstrated that any complex organ existed,
which could not possibly have been formed by numerous,
successive, slight modifications, my theory would absolutely break down. But I can find no such case. (Darwin,
*Origin*, 171)

There are plenty of such cases. How about hair for example? Can
a worm form it, thus turning into a caterpillar? What kind of hair
would that be and from which part of the body would it come? Let's
also take a look at internal organs, the brain in particular. Darwin did
not prove that natural selection changes the structure of the brain. If
his theory is correctly interpreted, it implies that no changes occurred

in the brain while a species was undergoing a plethora of changes regarding its bodily structure because the size, weight, and shape of the brain are immaterial to the survival of a species; only external organs are important. This leads to the ridiculous conclusion that, for example, the bird and the monkey have the same type of brain.

> Again, two distinct organs, or the same organ under two very different forms, may simultaneously perform in the same individual the same function, and this is an extremely important means of transition: to give one instance,—there are fish with gills or branchiae that breathe the air dissolved in the water, at the same time that they breathe free air in their swim—bladders, this latter organ being divided by highly vascular partitions and having a ductus pneumaticus for the supply of air. (Darwin, *Origin*, 172)

Actually, Darwin made a suggestion about where to look for transitional species. Unfortunately for the evolutionists, this fish is known to have been around for millions of years and yet no transitional species have been found in its habitat. This time it's not like looking for a needle in a haystack—evolutionists know the place and time, yet the geological structures reveal nothing.

> According to this view it may be inferred that all vertebrate animals with true lungs are descended by ordinary generation from an ancient and unknown prototype, which was furnished with a floating apparatus or swimbladder. (Darwin, *Origin*, 173)

This purported transition raises two objections. First, if a creature is versatile enough to live both in the sea and on dry land, why would it abandon half of its food supply and move out of the water? Even if, as evolutionists propose, it was pushed out of the sea by predators, it could occasionally go back to the water and replenish its food supply and then return to the dry land to nest. It is not clear why predators on land wouldn't push such a creature back to the sea. Maybe the creature was so big and strong that no land predator could touch it. But then the sea predators wouldn't touch it either, so it could have had the best of both worlds.

This brings me to the second objection. This ancient but unknown prototype was, presumably, a fish. While it may be possible to imagine a transition from a fish to a reptile, a transition from a fish to a mammal is an entirely different matter because two very different biochemistries are involved. The mammal's blood is warm because of very intricate biochemical reactions occurring in its body; because of their complexity reptile blood cannot duplicate such reactions. Darwin is, in effect, suggesting that a change in the shape or atrophy of an organ causes changes in the body's biochemistry. This is akin to saying that when coal is burned in a two-chamber furnace, it undergoes different chemical reactions from the ones occurring in a one-chamber furnace.

> The electric organs of fishes offer another case of special difficulty, for it is impossible to conceive by what steps these wondrous organs have been produced. But this is not surprising, for we do not even know of what use they are. In the Gymnotus and Torpedo they no doubt serve as powerful means of defense, and perhaps for securing prey. (Darwin, *Origin*, 174)

Actually, we know how these organs are used in at least some species, as Darwin remarked, so there is no mystery here. The mystery lies in development, as Darwin grudgingly admits. Is it possible to produce such an organ? Technologically challenged biologists, or naturalists, do not realize that this is an impossible engineering feat: the electric organ would have to evolve in isolation from the rest of the creature's parts, otherwise it would kill itself by sending an electric discharge throughout its own body. In order to prevent this from happening, the body would have to cut off the flow of blood to the electric organ because the blood is a good conductor of electric currents. But, as everyone knows, without blood flow the organ would be dead. There is only one way to produce an electric fish, namely, as a complete package all at once—which leads back to Creationism.

> The foregoing remarks lead me to say a few words on the protest lately made by some naturalists, against the utilitarian doctrine that every detail of structure has been produced for the good of its possessor. They believe that many structures have been created for the sake of beauty, to delight man or the Creator (but this latter point is beyond the scope scientific discussion), or for the sake of mere variety, a view

already discussed. Such doctrines, if true, would be absolutely fatal to my theory. (Darwin, *Origin*, 184)

Were animals created for the sake of their beauty? Beauty is in the eye of the beholder; some people find pigs extremely attractive and even keep them as pets while others don't, so this is all subjective.

Were animals created for the sake of variety? For some, 10,000 species is quite a large number, but others are never satisfied with any number of species. They want more. So this is subjective, too.

This is so typical of evolutionists—they pick on theories that are weaker than their brainchild, discredit them, and then declare victory, just as Darwin did. But what about stronger contenders? Or what about the *strongest* contender—my book? Can evolutionists prove that my criticism of their theory lacks logical foundation? This is a direct challenge, of course, but I'm sure they won't take on a fight they cannot win.

# 23 : Sharon Stone's Basic Instinct

> I will not attempt any definition of instinct. It would be easy
> to show that several distinct mental actions are commonly
> embraced by this term; but every one understands what is
> meant, when it is said that instinct impels the cuckoo to
> migrate and to lay her eggs in other birds' nests. (Darwin,
> *Origin*, 228)

Perhaps this is not an instinct at all; perhaps the cuckoo is smart
enough to watch and learn. When the phrase "mental actions" is used,
a reference is made to the thinking process and not to an instinct (and
it is important that we maintain a distinction between the thinking
process and instinct). Darwin's refusal to define instinct is a serious
flaw. Every science begins with definitions; without them a theory is
so vague that it cannot be successfully applied.

> Frederick Cuvier and several of the older metaphysicians
> have compared instinct with habit. This comparison gives, I
> think, an accurate notion of the frame of mind under which
> an instinctive action is performed, but not necessarily of its
> origin. How unconsciously many habitual actions are per-
> formed, indeed not rarely in direct opposition to our con-
> scious will yet they may be modified by the will or reason.
> (Darwin, *Origin*, 228)

By this definition, the word *habit* is synonymous with the word
*instinct*, which is not true according to the norms of the English
language, or any other language. Darwin mistakenly suggests that
instincts act in opposition to rational thinking, but if that were true,
an instinct would have been detrimental to the individual's survival.
Even the metaphysicians know that this is absurd. If an instinct can be
modified, then its original form is unknown and nothing can be said
about its influence on the individual's behavior.

> Habits easily become associated with other habits, with cer-
> tain periods of time, and states of the body. When once ac-

quired, they often remain constant throughout life. Several other points of resemblance between instincts and habits could be pointed out. As in repeating a well-known song, so in instincts, one action follows another by a sort of rhythm; if a person be interrupted in a song, or in repeating anything by rote, he is generally forced to go back to recover the habitual train of thought . . . (Darwin, *Origin*, 228)

Habits beget habits? This is something new. No one except for Darwin has been able to observe such a strange phenomenon. This idea is essential to a presentation of the instinct as something *acquired* by species, not something they were born with. Once again, Darwin's imagination has run wild. When Darwin gets high on evolution, he openly contradicts himself—in the preceding paragraph he wrote that habits, or instincts, could be overridden by the individual's will. This apparent contradiction is easy to explain: Darwin needs to show that, simultaneously, old instincts could be overridden and new, unshakable ones could be brought in as replacements. The evolutionary theory is so flexible that it can be bent and twisted to any degree. Any other scientific theory would have died long ago after such acrobatics, but this one is too stubborn to go away.

If we suppose any habitual action to become inherited—and it can be shown that this does sometimes happen—then the resemblance between what originally was a habit and an instinct becomes so close as not to be distinguished. (Darwin, *Origin*, 229)

Where did this come from? A great deal of research data proves otherwise: habits are never passed on to the next generation. And there is a good reason for that—such transference would require a change in genetic code without mutation, which is impossible. Granted, Darwin did not have genetic data, but even in his time it was known that habits die with the death of an individual.

It will be universally admitted that instincts are as important as corporeal structures for the welfare of each species, under its present conditions of life. Under changed conditions of life, it is at least possible that slight modifications of instinct might be profitable to a species; and if it can be shown that instincts do vary ever so little, then I can see no difficulty in natural selection preserving and continually accumulating variations of instinct to any extent that was profitable. (Darwin, *Origin*, 229)

There is a fundamental difference between instincts and corporeal structures: structures can be subjected to physical measurements while instincts are nonmeasurable. Without attaching a numerical value to an unknown, ill-defined quality it is impossible to tell how important, or unimportant, it is. Without a way to measure instincts, it is also impossible to tell if an accumulation of instincts can occur. Darwin's phrase "accumulated variations" essentially represents nothing; it's a meaningless combination of words.

> But I believe that the effects of habit are in many cases of subordinate importance to the effects of the natural selection of what may be called spontaneous variations of instincts;—that is of variations produced by the same unknown causes which produce slight deviations of bodily structure. (Darwin, *Origin*, 229)

Does natural selection produce changes in habits? Not according to Darwin. He believes that they both are produced by the same unknown causes. This means that the formation of instincts is not affected by natural selection because that is a *known* cause. Does natural selection produce changes in habits? Not according to Darwin. He believes that both habits and instincts are produced by the same *unknown* causes, which disqualifies natural selection from being one of the causes of instinct formation. This is the only possible interpretation of Darwin's definitions of instincts and habits—that there is no relation between natural selection and instincts. What is even more troubling for the evolutionists is that Darwin's theory does not exclude the possibility of animal behavior depending on instincts only, as modern scientific data suggests. As usual, Darwin began with contradictory definitions of instincts that lead to contradictory inferences about the role of natural selection.

> Again, as in the case of corporeal structure, and conformably to my theory, the instinct of each species is good for itself, but has never, as far as we can judge, been produced for the exclusive good of others. (Darwin, *Origin*, 229)

There is a multitude of bodily structures, but, according to all biologists, there are few basic instincts. This means that a large number of species share the same instincts. Why is this so important? Because this would imply that both predator and prey may share the same instincts, which would provide neither of them a clear advantage, and therefore neither of them would make evolutionary progress, which, in turn, implies that there is no natural selection.

> I can only assert that instincts certainly do vary—for instance, the migratory instinct, both in extent and direction, and in its total loss. So it is with the nest of birds, which vary partly in dependence on the situations chosen; and on the nature and temperature of the country inhabited, but often from causes wholly unknown to us: Audubon has given several remarkable cases of differences in the nests of the same species in the northern and southern United States. (Darwin, *Origin*, 231)

There are certain rules of logical inference to be followed if one wishes to arrive at a true conclusion; Darwin completely ignores them and his conclusion is simply laughable. To start with, is it the egg or the hen that came first? If the loss of migratory instinct is purportedly caused by natural selection, then one has to prove that natural selection exists without referring to the loss of migratory instinct as supporting evidence. If evolutionary progress is caused by changes in instinct, one has to prove that such changes actually took place without using natural selection as supporting evidence. Darwin does neither. He uses what is known as circular logic (this type of "logic" uses proposition A to show the truth of proposition B, and then uses proposition B to demonstrate the truth of proposition A, so both propositions appear to be correct).

Did the audubon change its instinct when it moved from northern to southern U.S., or was it a display of two sides of multiple instinct depending on geographical location? There is no way of telling which guess is correct because there is no scientific data to support either claim. But Darwin does not care about the data; he makes completely arbitrary statements that come out of thin air.

> How strongly these domestic instincts, habits, and dispositions are inherited, and how curiously they become mingled, is well shown when different breeds of dogs are crossed . . . These domestic instincts, when thus tested by crossing, resemble natural instincts, which in a like manner

become curiously blended together, and for a long period
exhibit traces of the instincts of either parent . . . (Darwin,
*Origin*, 233)

Is it fair to compare instincts to characteristics acquired due to
the crossing of species? It takes countless generations to develop an
instinct, as Darwin suggested, but the results of a cross can be seen
in the very next generation; besides, hybrids might have plenty of
undesirable characteristics because they were not subjected to natural
selection. Therefore, instincts and hybridization are completely unrelated phenomena. Evolutionists, starting with Darwin, were using
hybridization to prove that natural selection leads to the formation
of new instincts. But this total absence of logic renders their "proof"
akin to an old theory that earthquakes are caused by abrupt changes
in air temperature.

> Natural instincts are lost under domestication: a remarkable
> instance of this is seen in those breeds of fowls which very
> rarely or never become "broody," that is, never wish to sit on
> their eggs. Familiarity alone prevents our seeing how largely
> and how permanently the minds of our domestic animals
> have been modified. (Darwin, *Origin*, 234)

How about a hen sitting on her eggs, an idyllic picture of rural
life? There were no incubators in Darwin's day, and he avoids the
question, Who sat on the damn eggs? Perhaps it was Darwin himself!
What about a cat hunting mice? Is the hunting instinct a natural one?
Darwin tries to convince the audience and himself that the presence
of natural instincts separates wild animals from domestic ones, but in
reality this is not the case.

> Domestic instincts are sometimes spoken of as actions
> which have become inherited solely from long-continued
> and compulsory habit; but this is not true. No one would
> ever have thought of teaching, or probably could have
> taught, the tumbler-pigeon to tumble,—an action which,
> as I have witnessed, is performed by young birds, that have
> never seen a pigeon tumble. We may believe that some one
> pigeon showed a slight tendency to this strange habit, and
> that the long-continued selection of the best individuals in
> successive generations made tumblers what they now are . . .
> (Darwin, *Origin*, 233)

First Darwin suggests that instincts are compulsory habits (see the previous chapter), then he says that this is not true—a strange tumbling habit was passed on by one pigeon to future generations. He changes the word *compulsory* to the word *strange* and insists that this is a completely different case. Darwin needed this change because the evolutionary theory is completely incapable of explaining the origin of the tumbling phenomenon. The trick itself is useless. It hardly contributes anything to the survival of the species; nontumblers somehow managed to survive without it. Perhaps this behavior is not an instinct but rather some sort of play, similar to the way little kittens play with small objects. According to the evolutionary theory, animals are incapable of having fun because their behavior is determined by instincts and nothing else. If this line of reasoning is followed to its logical conclusion, it can be inferred that people are not supposed to have a sense of humor because their behavior is predetermined by instincts, albeit more sophisticated ones. The existence of senses of humor poses a challenge to the notion of survival of the fittest because when people laugh their muscles become relaxed, so they become more prone to accidents, and their chances of survival diminish.

> *Instincts of the Cuckoo* . . . And analogy would lead us to believe, that the young thus reared would be apt to follow by inheritance the occasional and aberrant habit of their mother, and in their turn would be apt to lay their eggs in other birds' nests, and thus be more successful in rearing their young. (Darwin, *Origin*, 235)

Supposedly, instincts bring improvement to a species. Is the cuckoo improved in any way by this particular instinct? "Illegitimate" cuckoo chicks raised by other birds do not experience any improvement in such qualities as the ability to procure food, the ability to evade predators, physical strength, etc. Some biologists have argued that by laying their eggs in other birds' nests, cuckoos increase the number of their offspring, thus increasing the chances of survival for the whole species. However, there is no evidence that cuckoos lay eggs more frequently than other bird species. On the contrary, cuckoos produce fewer eggs than other birds because they lay no more than one egg in a neighbor's nest in order to avoid competition for food between their offspring. In fact, this cuckoo policy puts the cuckoo at a disadvantage.

Darwin suggests something else that practically all biologists strongly disagree with—that acquired behavior can be inherited by the next generation (which is, for Darwin, the only way to explain how the cuckoo got its instinct). But this is akin to saying that, for example, the offspring of a dog who was trained to fetch small objects will copy her behavior *without* any training. Everyone knows how absurd this suggestion is.

* * *

> This ant is absolutely dependent on its slaves; without their aid, the species would certainly become extinct in a single year. The males and fertile female do no work of any kind, and the workers or sterile females, though most energetic and courageous in capturing slaves, do no other work. They are incapable of making their own nests, or of feeding their own larvae. When the old nest is found inconvenient, and they have to migrate, it is the slaves which determine the migration, and actually carry their masters in their jaws. . . . By what steps the instinct of F. sanguinea originated I will not pretend to conjecture. (Darwin, *Origin*, 239)

No steps could have led to this instinct because it is self-destructive for the masters. It certainly looks in this instance as though natural selection reversed itself and came to the rescue of the most unfit individuals. Then natural selection did an even stranger thing—it took the most capable individuals and turned them into slaves with a dramatically reduced ability to survive because now they have to feed not only themselves but their masters as well. The timing of the instinct is unbelievably precise: two species developed their instincts simultaneously. How great is the chance of that? Probably zero. Darwin definitely succeeded in choosing the instincts for his study that are certain to destroy the evolutionary theory!

> We hear from mathematicians that bees have practically solved a recondite problem, and have made their cells of the proper shape to hold the greatest possible amount of honey, with the least possible consumption of precious wax in their construction. (Darwin, *Origin*, 242)

To develop such perfect instinct step-by-step, bees must have had some memory of previous attempts at building honeycomb, otherwise no improvement would have been possible. But such outstanding memory would require a fairly high degree of intellect. Now we have bees with PhDs in mathematics whose intellectual capability exceeds that of Darwin!

It would be interesting to see how Darwin's study of instincts translates into the world of men. All men are supposed to have the same instincts, just like all migratory birds have their migration instinct. If this were true, all men would drool when they saw Sharon Stone displaying a certain body part. Some men did indeed drool, others felt offended, and a small number of men, including this author, found the movie so boring that they turned the TV off without seeing the ending.

What is a correct definition of instinct? Certainly it cannot be defined as a mental state because mental states are not measurable ("mental state" is a figure of speech; it should not be taken literally). An instinct is defined through physical actions involved in its fulfillment. For example, a beaver's instinct should be defined as a series of steps takes by a beaver to build a dam; instinct serves as a kind of construction manual. Any other definition of instinct involves reference to a mental state, which would mean that it was a metaphysical quality. It is impossible to tell, based on the observation of an instinct, whether it evolved step-by-step or was present from the beginning in the very same form; all the observer has is a sequence of current events and nothing more.

> The fertility of varieties, that is of the forms known or believed to be descended from common parents, when crossed, and likewise the fertility of their mongrel offspring, is, with reference to my theory, of equal importance with the sterility of species; for it seems to make a broad and clear distinction between varieties and species . . .
>
> It is certain, on the one hand, that the sterility of various species when crossed is so different in degree and graduates away so insensibly, and, on the other hand, that the fertility of pure species is so easily affected by various circumstances, that for all practical purposes it is most difficult to say where perfect fertility ends and sterility begins. (Darwin, *Origin*, 259)

In other words, it is impossible to define clearly what fertility, sterility and pure species are, but then it is also impossible to make a broad and clear distinction between varieties and species. The whole classification breaks down and the whole notion of a common ancestor becomes nothing more than a play of words that reflects a total absence of concept. Darwin tries to clarify this subject because it's essential to the survival of the evolutionary theory, but this is an impossible task because the evolutionary theory *can't* survive.

> Hybrids are seldom raised by experimentations in great numbers . . . if left to themselves, will generally be fertilized during each generation by pollen from the same flower; and this would probably be injurious to their fertility, already lessened by their hybrid origin. (Darwin, *Origin*, 260)

Probably, but not certainly—a definitive conclusion is not possible because experimentation did not produce a sufficient number of hybrids to make valid observations. Does it make sense to draw a conclusion based on a small number of experiments? A majority of statisticians would agree that this is *not* a good idea.

> He [W. Herbert] is as empatic in his conclusion that some hybrids are perfectly fertile—as fertile as the pure parent-species—as are Kölreuter and Gärtner that some degree of sterility between distinct species is a universal law of nature. (Darwin, *Origin*, 265)

Some species produce sterile hybrids. One does not have to have an advanced degree in biology to know that. But what does it mean? Biologists believe that the ability to produce hybrids, sterile or fertile, indicates common ancestors. But all organisms have, according to the same biologists, a common ancestor, namely, the original cell. This means that all species should be able to produce hybrids. For some scientists the fertility of hybrids is an interesting topic, but it does not add anything substantial to the evolutionary theory or to creationism.

> I have found it difficult, when looking at any two species, to avoid picturing to myself forms *directly* intermediate between them. But this is a wholly false view; we should always look for forms intermediate between each species and a common but unknown progenitor; and the progeni-

tor will generally have differed in some respects from all its modified descendants. (Darwin, *Origin*, 288)

What about the progenitor's progenitor? Then the progenitor becomes an intermediate species and the search for it is meaningless because there is no progenitor, unless we go all the way back to the original cell. But then the notion of multiple progenitors is useless. Apparently, Darwin is incapable of making a simple logical deduction. How would we know that there are intermediate forms? The only way to prove that certain species have a common progenitor is to find him. But the progenitor is unknown and the whole chain of Darwin's conjectures that are based upon his existence leads to nowhere.

> By the theory of natural selection all living species have been connected with the parent-species of each genus, by differences not greater than we see between the natural and domestic varieties of the same species at the present day; and these parent-species, now generally extinct, have in their turn been similarly connected with more ancient forms; and so on backwards, always converging to the common ancestor of each great class. (Darwin, *Origin*, 289)

Assuming that domestic species were evolving at the same rate as natural ones, how difficult would it be to find intermediate species? The geographical locations of major ancient cities are well-known, so a series of geological excavations would be sufficient to unearth extinct domestic species. Yet why has no one been able to find them?

According to Darwin and his followers, the classification of living beings into a genus is done on the basis of the existence of their progenitors. But the progenitors have their own progenitors, and so on. All parental forms lead to the original cell, which means that subdivision into species is impossible because they all have a common progenitor. What if there was more than one original cell as some biologists have suggested? Then there are two distinct possibilities:

a) *All original cells were created simultaneously.*

That would be some sort of miracle, and according to modern evolutionists this is not what happened.

b) *The original cells came into existence in different epochs.*

If this is the case then there is a nagging question—why has not a single original cell appeared in our epoch?

> [B]ut if we confine our attention to any one [geological] formation, it becomes much more difficult to understand why we do not therein find closely graduated varieties between the allied species which lived at its commencement and at its close. . . .
>
> I am aware that two paleontologists, whose opinions are worthy of much deference, namely Bronn and Woodward, have concluded that the average duration of each formation is twice or thrice as long as the average duration of specific forms. (Darwin, *Origin*, 298)

As the above paragraph shows, objections were raised a long time ago and today, as before, evolutionists fail to explain why not a single geological formation contains intermediate species. There were attempts to prove that intermediate forms stretch beyond a single formation, but then the number of formations becomes too small to provide enough time for evolution of species. The math is simple and it would be unnecessary to present it here. Anyone with a high school diploma can do this type of calculation. Darwin, however, was a bad high school pupil; even the simplest of calculations was too hard for him.

Darwin tried to compensate for lack of mathematical skills with geological considerations, as the chapter in his book titled "Absence of Intermediate Varieties" shows. But he is not a geologist either, so he came up with all kinds of ridiculous explanations.

> The abrupt manner in which whole groups of species suddenly appear in certain formations, has been urged by several paleontologists . . . as a fatal objection to the belief in the transmutation of species. (Darwin, *Origin*, 305)

The objection is truly fatal, although Darwin does not want to accept it. This data has been such a pain for evolutionists ever since it was discovered that several modern biologists proposed that there were sharp transitions between species. Others objected because this is a genetic impossibility. The whole story shows how desperate the evolutionists have become.

> To the question why we do not find rich fossiliferous depos-
> its belonging to these assumed earliest periods prior to the
> Cambrian system, I can give no satisfactory cause. (Darwin,
> *Origin*, 309)

A scientific theory might not have the answers to all questions
and still remain a viable theory, unless the unanswered questions
threatened its overall correctness. The above question is of fundamental importance to the evolutionary theory and yet it remains unanswered even now.

> The productions of the land seem to have changed at a
> quicker rate than those of the sea, of which a striking in-
> stance has been observed in Switzerland. There is some rea-
> son to believe that organisms high in the scale, change more
> quickly than those that are low: though there are exceptions
> to this rule. (Darwin, *Origin*, 313)

When intermediate forms are missing, it is impossible to determine the rate of change for both earth and sea creatures. All Darwin's
suggestions are wild conjectures so characteristic of the evolutionary
theory. The scale itself is not defined, so the whole discussion of high
and low organisms is unrealistic.

Is the ape higher on the scale than the sea horse? Both are equally
well adapted to their environments, so the adaptability factor puts
them on the same level. What other factors are left? The monkey
could be trained to perform in a circus while the sea horse could not.
But they are both wild animals that have not been subjected to domestication by natural selection, so this factor does not mean much.
Other than these two factors, the evolutionary theory offers no basis
for comparison, which means that these two organisms are evolutionarily equal. If we adopt the creationist point of view, they are clearly
not equal because their relationships to humans come into play—this
relation being nonexistent from the standpoint of the evolutionary
theory.

> We can clearly understand why a species when once lost
> should never reappear, even if the very same conditions of
> life, organic and inorganic, should recur. For though the
> offspring of one species might be adapted (and no doubt
> this has occurred in innumerable instances) to fill the place
> of another species in the economy of nature, and thus sup-

> plant it; yet the two forms—the old and the new—would
> not be identically the same; for both would almost certainly
> inherit different characters from their distant progenitors;
> and organisms already differing would vary in a different
> manner. (Darwin, *Origin*, 315)

They vary in a different manner, all right, but conditions have
returned to what they once were, and the natural selection, if it exists, will return species to the state they were in before the conditions
started to change. This is how natural selection is supposed to work,
isn't it? Otherwise, this conclusion is unavoidable: natural selection
does not create the best possible forms.

All scientific theories, except for the evolutionary theory, are
based on the premise that uniformity in conditions implies uniformity in results, otherwise no scientific development is possible. Darwin
probably knew that, but he tried to rebuff the critics who saw weakness in the evolutionary theory.

> A group, when it has once disappeared, never reappears; that
> is, its existence, as long as it lasts, is continuous. I am aware
> that there are some apparent exceptions to this rule, but the
> exceptions are surprisingly few. . . . (Darwin, *Origin*, 315)

Previously, Darwin wrote about the scarcity of geological records; it may be that there are more exceptions to this rule, and it
may be that there is no such rule at all. But then evolution goes on in
circles, and the world, by this line of reasoning, should be as it was in
the times of dinosaurs. Perhaps there is a better explanation: there is
no such thing as natural selection.

> We have seen in the last chapter that whole groups of species
> sometimes falsely appear to have been abruptly developed;
> and I have attempted to give an explanation of this fact,
> which if true would be fatal to my views. But such cases
> are certainly exceptional; the general rule being a gradual
> increase in number, until the group reaches its maximum,
> and then, sooner or later, a gradual decrease. (Darwin,
> *Origin*, 316)

How can a species increase in number if its members undergo
transmutation? Transmutation means that new species are developed, so the original species never has an opportunity to reach its

maximum. Even if they increase until the maximum is reached, why would their numbers decrease after that? In the beginning of his book Darwin introduced the concept of checks that keep a population at a stable level, with only small variations. What happens after they drop in number? Do they become extinct or grow until the maximum is reached? Perhaps they become abruptly developed and who knows what happens next. Everything in Darwin's theory becomes completely arbitrary. There is only one consistency—Darwin is constantly inconsistent in his estimates.

> The old notion of all the inhabitants of the earth having been swept away by catastrophes at successive periods is very generally given up . . . (Darwin, *Origin*, 316)

Practically every biologist accepts the fact that the extinction of the dinosaurs was caused by abrupt climatic changes. Darwin's assertion of gradual changes is incorrect; catastrophic events spell out catastrophe for the evolutionary theory.

> There is reason to believe that the extinction of a whole group of species is generally a slower process than their production. (Darwin, *Origin*, 317)

When intermediate forms are nowhere in sight, it is impossible to tell whether the production or the extinction is faster. Operating on the assumption that the evolutionary theory is correct, one comes to the conclusion that the production is faster, but the assumption is still unproven due to the lack of paleontological data. The only way out of this quandary is to use circular logic, which is itself a logical dud.

> If we ask ourselves why this or that species is rare, we answer that something is unfavourable in its conditions of life, but what that something is we can hardly ever tell. (Darwin, *Origin*, 317)

The terms *rare* and *extinct* are not synonyms. A small number of individuals do still remain in the former case. According to the evolutionary theory, those are the fittest individuals. Being the most adaptable, they should give rise to the most vibrant generations, multiply and prosper. But according to the principle of natural selection, then, they should have never become rare to begin with. Either this principle is incorrect or there are no rare species. "I am not a rare species," says the gorilla; Darwinists created the Endangered Species List because they want more grants to study evolution.

Time has come to part with Darwin's *The Origin of Species* book, with its endless repetitions of the same themes. I have covered 320 of 450 pages of his book—going any further would risk boring the reader to death.

Certain things have changed since the passing of Darwin—new genetic material was added to the evolutionary theory and so was the concept of random mutations. But one thing remained unchanged, that being the cornerstone of the theory of evolution, the principle of natural selection. Without this principle the theory is dead; not even random mutations and new classifications of species can save it.

My goal in the last three chapters, which are devoted to criticism of Darwinism, is to show that natural selection is a myth. Nothing of this sort can possibly exist in the nature. Have I achieved my goal? Let the reader be the judge!

# 24 : Systematics

Ernst Mayr was hailed by evolutionists as one of the great modern-izers of the evolutionary theory. His book *Systematics and the Origin of Species* is considered to be a major contribution to the field. But the book is full of egregious errors and unproven propositions that make it absolutely useless.

I'll start my critical review of the book with chapter I.

Mayr begins the first chapter with the admission that there are disagreements between animal taxonomists, plant taxonomists, zoolo-gists, parasitologists, etc., with regard to how to answer the following questions, among others: Are the systematic categories natural? What is a species? etc. Other than stating the fact that there are differences of opinions, Mayr fails to explain what causes them; apparently, he lacks knowledge of the basics of classification.

Every classification is made for the practical purpose of con-veying information in the shortest possible form. There is no other purpose in assigning objects, or living beings, to classes. Zoologists and plant taxonomists, for example, deal with very different kinds of organisms, so their ideas of convenient classification disagree. This does not mean that one group has a better classification than the other; it only proves that practicality is relative. The words *convenient* and *natural* are not synonyms; it would be correct to say that natural categories do not exist.

Mayr believes that the old classification based on a morphologi-cal definition of species (i.e., one in which species were categorized based on similarities in physical appearance) was replaced by a better classification that was based on ecological, genetic, geographical, and physiological factors. On the surface this seems like a considerable improvement. However, a closer look reveals plenty of inconsistencies and flaws. Take a genetic factor, for example. Mayr knows that only 0.05% of all animals have their genetic makeup completely mapped out by geneticists; this glaring lack of data makes any genetic classifi-cation a joke. But even if the genetic makeup of all animals is known, there is an insurmountable obstacle to such classification, which can be summed up in one question: how is the genetic distance between two species defined? There were attempts to define this distance as a number of genes common to two species. This definition is completely wrong because it's not separate genes that matter most but rather combinations of them. To illustrate this difference let's consider a chemical example: both $H_2O_2$ and $H_2O$ contain hydrogen and oxygen, but their properties are so different (you cannot intake $H_2O_2$ without the risk of seriously injuring yourself) that it prevents them from falling into the same category of substances no matter what classification scheme is used. There is another objection to this definition of genetic distance as well: it puts certain animals closer to inanimate objects such as trees than to other animals.

Physiological similarities cannot be taken as a basis of classification either. As scientists experimenting with animals in order to find cures for human illnesses know, there is not a single animal species best suited for all kinds of treatment. Consequently, there are several types of laboratory species. This means that "physiological distance" is impossible to define.

How about ecological factors? As the botanists know, there are more ecological factors than there are species, so any classification based on regional ecology is impractical. This shows, contrary to Mayr's belief, that systematics is not a full-fledged science but is rather a dud.

Is it possible to conclude with a high degree of certainty that two similar forms are the same species or distinct species? Mayr believes that proper classification allows one to make a determination. However, since all classifications are arbitrary, no such determination is possible. It is also impossible to determine whether a similarity between two species is due to similar habits or a close phylogenetic relationship (having a common progenitor) because there are no criteria. However, if we adopt the view that intergradation, or evolutionary development, does not exist, the only remaining option is hybridization.

Is it possible to use a classification of species as a proof of the evolutionary theory? Mayr clearly believes so because he cannot accept the view that genera and families are not objectively real but only artificial creations of the taxonomist.

A failure to see the obvious casts doubt on Mayr's credibility as a taxonomist. "One of the principal advantages of the binary system is that it permits the omission from the description of all characters common to the genus, family and higher categories. To include these 'higher' characters in the species diagnosis would only obscure the true diagnostic features" (Mayr, *Systematics and the Origin of Species*, 13).

In any scientific branch this would have been seen as the most ridiculous proposition: a classification without defining characteristics. But it works for many biologists unfamiliar with the basic rules of science.

There is another sign of trouble in the opening chapter where Mayr suggests that a description of a specimen should include its maximum, minimum, and mean values. While the mean is of great importance, the minimum and maximum values are not important at all. Instead, the variance should be included in the description, as every statistician knows. Mayr's advice shows that he is unfamiliar with the basics of mathematical statistics, which is a serious shortcoming that led to an incorrect hypothesis about certain small, isolated populations being the source of new species (I'll return to this topic soon).

Apparently, Mayr believes that the presence of a zoological classification shows that the evolution of species did occur. But this is a great misunderstanding of the role of a classification—it does not prove or disprove that certain events occurred in the past, or will come to be in the future, but rather simply reflects the world as it is today. Take the periodic system of elements, for example. It tells nothing about how all known elements were created after the big bang; it simply reflects their current properties. It would be too much to expect from a classification system to provide a description of past events because it was not designed for this purpose, as all scientists, except for biologists, know.

In Chapter II Mayr runs into trouble distinguishing essential characters from nonessential ones. For example, he states that when related insect genera are compared, the presence of two extra bristles on the thorax may be the least important characteristic, although it is the most easily recognizable one. This confusion is easy to explain because there are no rules that separate important characters from unimportant ones. Everything is left to the discretion of a taxonomist. This discussion seems academic, but it goes to the heart of the matter when biologists assert that their classifications support the evolutionary theory. But their categories are arbitrary to such a degree that in many cases they disagree with each other.

After making a remark about physiological differences in a diagnosis being undesirable because they come from dead specimens, on the next page Mayr suggests that they should be used as further refinement when morphological differences are not pronounced. Which approach is correct? Mayr did not clarify his position because he knows very little about physiology, which is not his field. The same is true for other taxonomists, so their classification schemes fall back to the times of Darwin. The sad truth for biologists is that, despite the discoveries of new species, no additional proofs of the evolutionary theory have been presented. Well, maybe genetics can help. Since very little has been done in this field as well, except for the mapping of the human genome, Mayr implicitly suggests that the number of genes of a species can be used as a basis of classification. According to him, those organisms higher on the evolutionary scale have the highest amounts of genes. Perhaps Mayr did not know that the highest-ranking organism, a man, does not possess the largest amount of genes. There are several "primitive" organisms with larger numbers of genes, as was recently discovered. This not only means that genetic classification is a wishful thinking, but also that evolution did not end with man but went on further to more complex, albeit not very intelligent, beings.

Are morphological characters correlated with physiological and biological characters? Mayr believes they are; he even cites several taxonomic publications to support his point of view. At the same time he admits that taxonomists disagree about which species are good and which are not. The correlation goes out the window! Jokes aside, there is a deeper reason for not having a correlation. Suppose that a group of geneticists have mapped the genome of a species, then the other group is presented with the findings without being told which animal was investigated. Can they guess, based on genetic information, the physical characteristics of the animal? Geneticists would agree that this is impossible. But this also proves that there is no correlation between morphological and genetic characters. The same conclusion holds true for physiological characters; knowledge of sizes and positions of internal organs and chemical reactions occurring in them does not reveal much about morphological characters. The absence of correlation is a very important fact precluding any notion of evolutionary stages. But if knowledge of something is unobtainable in principle, then that something can essentially be declared false, as every real scientist knows. This makes the evolutionary theory a prominent example of a faulty theory.

Mayr proposes using individual variation (variation within a population) and group variation (variation between different populations within a species) as a basis of classification; at the same time he admits that the population concept is hard to define. It's no wonder why—there is disagreement between taxonomists on what should be called species. Because of this disagreement, individual and group variations become indistinguishable. Such flexibility leads to the assignment of different progenitors to a species by different teams of taxonomists. Their inability to determine progenitors with any certainty puts the whole evolutionary theory in doubt because it becomes unverifiable. Mayr tried to bypass this difficulty by declaring that a certain genetic similarity introduces the possibility of interbreeding and, because of that, could be used to define group variation. This is a rational suggestion. However, it is applicable to living organisms only; nothing regarding interbreeding can be inferred from a study of the DNA of extinct species, which confirms the impossibility of determining the progenitors and therefore of saving the evolutionary theory.

In chapter III Mayr goes into considerable detail showing how characteristics such as size, proportion, epidermal structure, coloration, etc., vary in different populations of a species depending on the geographical region (geographical variation). It is an interesting reading and, I suppose, makes zoologists warm and fuzzy inside. It is not clear, however, what geographic variations signify. There are no patterns, which makes them unanalyzable. In addition, as Mayr noted before, there is no agreement on what should be called a species. The unanalyzability of patterns and the absence of a universally accepted classification system make the evolutionary theory unverifiable. Kant and Hegel believed that certain unverifiable propositions are correct, but their philosophical systems are things of the past (though biologists still cling to them primarily because of limited mental capability). But positivists, whose ideas triumphed in a majority of the natural sciences, have demonstrated that unprovable propositions are false and should be discarded.

## Genetic Drift

Mayr's complete unfamiliarity with mathematical statistics leads to the belief that genetic drift may overcome the effect of sexual selection in well-isolated and small populations. Besides the fact that sexual selection does not exist, as I have already proved, genetic drift produces no appreciable effect on the offspring because, after fifteen generations, it lessens to the point where it becomes practically zero, as a great mathematician, R. Fisher, has shown. From a mathematical standpoint, the size and isolation of a population do not matter because the probability of a meeting of two individuals with the genetic make-up required to transmit desirable characteristic to the next generation is determined independently of the size of the population. This can be demonstrated as follows: suppose a small population consists of N species and a large population of M species. The number of individuals with advantageous characteristics in the small population is n; in the large one, m. The ratio $n/N$ signifies the probability of the right individuals meeting in the small population; $m/M$ in the large one. Since these populations constitute the same species, $n/N = m/M$ (the ratios are never exactly equal, but their difference is statistically insignificant). The equality holds unless there is a reason to believe that these populations are not uniform, which never happens in nature because there are no known factors that cause nonuniformity.

Here is a recap: the probability of the transmission of desired characteristics to future generations does not depend on the size of a population or its degree of isolation.

## New Classification

In Chapter V Mayr describes what he calls a "revolution in taxonomy": an old, static definition of species was replaced by a new definition based on geographic proximity.

> The basic idea is that the next relative to any species occurs, in a more or less disguised form, in a geographically adjacent area. (Mayr, *Systematics*, 119)

Not much of a revolution if you ask a nontaxonomist! However, it raises an interesting question: what is a geographically adjacent area? No matter how geographically adjacent areas are defined, moving from one to another can cover, excluding islands, a whole continent, so this is hardly an improvement of the old classification. Mayr needs this "improvement" to show that the alleged effects of evolution are such that several species living in adjacent areas are certain to have a common progenitor. But this new classification scheme without geographic borders is so amorphous that it does not prove anything.

Another of Mayr's definitions of species does not make much sense either:

> Species are groups of actually or potentially interbreeding natural populations, which are reproductively isolated from other subgroups. (Mayr, *Systematics*, 120)

It seems that Mayr knew the weakness of his definition when he suggested that it be left to a taxonomist to decide whether two particular species are potentially capable of interbreeding because in many cases direct experimental data is not available. The lack of experimental evidence shows that Mayr's definition of species is not universal at best, which could be a fatal shortcoming. Now, let's take a look at those species capable of interbreeding. Does their ability to interbreed prove that they have a common ancestor? First and foremost, there are some species, including a man, that do not interbreed with any other species. If Mayr is correct, these species do not have progenitors because there are no related species. Some evolutionists tried to explain the inability to interbreed by suggesting that related species became extinct. But this is hardly an explanation because in the past they were able to interbreed with closely related species, which, in turn, leads to the production of, for example, a race of man-like creatures living in the present time. Other evolutionists tried to circumvent this difficulty by suggesting that the hybrids were sterile. It's true that certain species produce infertile hybrids, but not the ones that came to be right after the branching. Otherwise we must conclude that from a genetics standpoint, the branching of a species is impossible to define because in some cases it produces fertile hybrids, in some cases sterile ones, and in still other cases no hybrids at all.

In chapter VI Mayr presents statistical data for the purpose of demonstrating that while a minority of the species is monotypic, the majority is polytypic (having several different types). Mayr offers two explanations of the discrepancy: monotypic species occupy such small geographic regions that there is no opportunity for developing variation, and there is some unknown reason for their being monotypic.

What prevents them from expanding their range? The other species could move into new regions, but not them. Perhaps they were kept in small regions by predators or due to lack of food in larger regions. Why, then, did the predators not venture into their region? Why is there a lack of food in surrounding regions while there is plenty in their region? Droughts that cause food shortages spread evenly, so their region should have also been affected. Is it climatic difference that confines them? But their regions are too small for there to be a sharp climatic difference, so the climate is not a factor either. There is no reasonable explanation, unless there is some unknown factor, as Mayr proposes. But "unknown reasons" put a scientific theory on the same level with astrology, which is not a science but prejudice. Actually, the factor *is* known, although Mayr does not want to admit it, because admitting it is equivalent to saying that the evolutionary theory is a fantasy.

## Evolution According to Mayr

> A new species develops if a population which has been geographically isolated from its parental species acquires during the period of isolation characters which promote or guarantee reproductive isolation when the external barriers break down. (Mayr, *Systematics*, 155)

Mayr spends considerable time trying to prove the above statement in chapter IV. However, the first thing that comes to mind is this: how do geographical isolations actually occur? Usually, evolutionists give the following answer: a small group of individuals could be separated from the main population by a river, for example; and there are many other ways of achieving a separation as well. And they expect everyone to be satisfied with this answer. But the rivers do not appear out of nowhere; if the members of a species could swim across the river once, they could do it on many occasions. But this example does not clarify anything. Let's take a look at other possibilities:

1) *The individuals swam across the river, but then it rose, preventing their return.*

That would not have been a big deal. The river would have receded in less than two months, allowing them to swim back and forth again.

2) *Living conditions were better on the other bank.*

It's true that hey would not return in this instance, but this is not a geographic isolation because they can go further. Besides, the remaining individuals can do the same because the river is not much of a barrier.

3) *The mountains separated them.*

Mountains are not much of a barrier either, so the same argument holds.

4) *There was a huge volcanic eruption that changed the landscape.*

This is a possibility, but such eruptions are things of the past. They occurred during the earth's formative period, long before any species came into existence.

5) *Somehow several individuals got to an island and decided to stay there.*

That could happen only if they could talk and make collective decisions, which is clearly not the case. Most likely they would have continued moving to other islands or returned to the mainland (not all at once but one by one).

6) *There was a strong earthquake and a newly formed canyon separated them.*

There is always a way around a canyon. Besides, this is a dry land so they could walk up and down the canyon.

There are no other types of geographic separations for earthbound creatures, which means, in light of my rebuttals above, that it did not happen. Also, it is true that there is no such geographic barrier that a bird could not overcome. If Mayr's theory is correct, there should not have been new bird species forming; somehow he overlooked this obvious conclusion.

Trying to add more credibility to his theory, Mayr enlisted the help of a fellow evolutionist:

> The two populations may differ little if any at the time of separation, but will drift even further apart, each carrying its subspecies with it. The accumulation of genetic, chro-

mosomal and cytoplasmic differences tend to lead in the course of ages to inter-sterility or hybrid sterility, making irrevocable the initial merely geographic or ecological isolation. (Wright, "Material Basis," 165–70)

Wright's theory is completely illogical. To start with, there are no barriers between subspecies, which means that they should not be forming. They should remain the same species. Assuming that subspecies are forming, they should be accumulating genetic, chromosomal, and whatever other differences there are too, thus raising their status to species. If their hybrids are sterile, then, according to Mayr, they are separate species right from the start, and no geographic isolation is needed to create them. If the hybrids are fertile, then all of them should mix, leading to the production of just one species without the subspecies. Both Mayr and Wright are incapable of logical deductions, but what else would you expect from evolutionists?

## Instantaneous Creation of Species

Mayr believed that in addition to geographic isolation, there exists another, equally important mechanism called instantaneous sympatric speciation, which plays a role in the production of species.

> The term instantaneous sympatric speciation means the production of a single individual (or the offspring of single mating) which is reproductively isolated from the species to which the parental stock belongs. Such an individual could be the potential ancestor of a new species. (Mayr, *Systematics*, 190)

That such a mechanism exists is a highly speculative hypothesis. Mayr does provide what he believes to be examples from the plant kingdom, but they could be disputed. What is certain is that no such thing was observed among the animals, and there is no genetic data to support this assertion.

There is a somewhat similar hypothesis that an individual could acquire a new, advantageous characteristic as a result of a mutation and could then spread it to the entire population. Although Mayr did not mention this hypothesis in his book, it would be relevant to conclude this chapter with a discussion of the mathematics behind the spread of a mutation.

Let's assume that a mutated individual with an advantageous characteristic occasionally appears in every population. Can he spread this characteristic to the entire population? R. Fisher showed that this is extremely unlikely (see Chapter XIII of this book). Several evolutionists suggested that natural selection changes things dramatically, concluding, then, that the possibility of transmission of new characteristics becomes very high. But their knowledge of mathematics is very limited; otherwise they would have known that natural selection makes things even worse, as will be demonstrated below.

To keep things simple and to avoid the use of mathematical statistics, we'll consider the most favorable case: a mutated individual was able to transmit his newly acquired characteristic to all his children. Let the total number of members of a population be N, and the number of improved individuals, m. Now, natural selection comes into play. Let "a" be the number of non-improved species eliminated by natural selection, and "b" be the number of improved species also eliminated by it. We assume that b > 0; this indicates that improved species are affected by natural selection, although to a lesser degree than non-improved ones ( b < a ). If we assume that the improved species are not affected by selection (b = 0), that would prove that further improvement is impossible and no new species will evolve, which is equivalent to saying that the evolutionary theory does not work.

Now we form the following ratios:

$Z_1 = m/N$ and $Z_2 = (m - b)/(N - a - b)$

where $Z_1$ is the percentage of new, improved individuals before natural selection took place, and $Z_2$, the percentage after natural selection. Let $Z = Z_2 - Z_1$

Everything depends on the sign of Z. If Z > 0, that would indicate that the percentage of new species is increasing, Z < 0 would indicate that the percentage of new species is decreasing, and Z = 0 would indicate that a case of new species is remaining at the same level. After performing a simple calculation we find that

$Z = (ma + mb - Nb)/(N - a - b)N$

N > 0 and N - a - b > 0, so we need to worry about the sign of the nominator only (the nominator is ma + mb - Nb)

N represents an order of magnitude of at least 1,000 while m represents an order of magnitude of 10, and a and b are even smaller. This shows that $Z < 0$ and, as we noted before, it means that the percentage of improved species is falling down because of natural selection. This simple example will illustrate the calculations:

Let $N = 1000$, $n = 10$, $a = 2$, $b = 1$.
Then $Z_1 = 1\%$ and $Z_2 = 0.3\%$

On this mathematical note we part with Mayr's book.

# 25 : Genetics and Man

Theodosius Dobzhansky is seen by many evolutionists to be the greatest defender and disseminator of the evolutionary theory. Perhaps Dobzhansky became famous because his books are easy to read and, unlike the books of other proponents of the evolutionary theory, they are quite entertaining. But entertainment aside, they lack logical foundation and the argument is extremely weak.

> Moreover, modern cosmology, the study of the universe, assumes that the atoms that exist today have had a history. One of the theories is that the atoms were formed from a primordial substance called the "ylem", and the inference is that they were formed in a tremendous explosion, which occurred supposedly more than 5 billion years ago. This stupendous event is, then, the first discernable date in the history of the universe and the beginning of *cosmic evolution*. After the formation of the atoms, cosmic evolution led to the concentration of the original cloud of atoms into galaxies. Our earth is an insignificantly small particle of the universe, yet we cannot be sure that life exists anywhere except on this small particle. The evolution of life, *biological evolution*, thus, to our knowledge, been enacted on earth alone. (Dobzhansky, *Evolution*, 1)

This is just one example of Dobzhansky's illogical constructs. To start with, nothing indicates that earth is the only cradle of life. There may be plenty of other planets suitable for life that scientists simply do not know about yet.

Cosmic evolution does not imply biological evolution, as Dobzhansky suggests. These are two completely unrelated topics, as every high school student can see. It would be a waste of the reader's time to go into unnecessary details to show the absence of a logical connection between cosmic evolution and existence of the original cell.

## Arbitrary Classifications

*Genetics and the Origin of Man* is Dobzhansky's most popular book. The book is so full of gross logical errors and defective deductions that only an evolutionist could love it.

The opening chapter in Dobzhansky's book begins with two contradictory statements:

> Classification is natural and not artificial, in so far as it reflects objectively ascertainable discontinuity of variation, and in so far as the dividing lines between the discrete clusters of living forms

and

> Biological classification is simultaneously a man-made system of pigeonholes devised for the pragmatic purpose of recording observations in a convenient manner, and as acknowledgement of the fact of organic discontinuity. (Dobzhansly, *Genetics*, 1)

So, is the classification natural or man-made? Dobzhansky believes that it is both, but that could only be true if just one classification system were to exist. As Mayr pointed out, there are numerous classification systems (see the previous chapter of this book); this clearly shows that Dobzhansky's assessment is wrong.

Dobzhansky also concluded that discontinuity of the variations among organisms proves that classification is not artificial; this implies that continuity would have proved that classification is man-made.

"Continuity" and "discontinuity" are subjective categories, as a majority of scientists know, because continuity does not actually exist. It is just a mathematical abstraction. This is easily explainable: every measurement is discrete because its value cannot be smaller than the resolution of the measuring apparatus; different apparatuses provide different degrees of precision, but none can measure infinitely small distances or intervals of time, which are the properties of continuous measurements. Continuity is a very useful abstraction that is used in every branch of science, but it is not one of the properties of the real world. Contrary to Dobzhansky's understanding of science, subjective categories cannot be used to prove or disprove any assertion.

No lion cub is ever born to a pair of cats, nor the converse ever observed. A species is, consequently, not merely a group and a category of classification. (Dobzhansky, *Genetics*, 3)

According to Dobzhansky, a species is defined as a group of individuals who sexually reproduce offspring that are in some way similar to them. From the logician's standpoint, this is a vicious circle: first species are defined as a group of similar organisms, then it is said that because of a similarity between the parents and children, they belong to the same species. In other words, a species is defined as a group of organisms with a certain parent-child relationship (resemblance), and at the same time this parent-child relationship is defined as a species. Just like any other evolutionist, Dobzhansky is walking blindly in circles.

To demonstrate how arbitrary the classifications are, let's take a look at another example, dogs and wolves. German shepherds and dachshunds cannot interbreed because of the difference in their sizes, but that does not preclude them from falling into the same category—dogs. German shepherds and wolves can interbreed, and yet they belong to different categories—dogs and wolves, respectively. This example should erase any doubt about the arbitrariness of classification.

## Deities & Extraterrestrials

The greatest achievement of biological science to date is the demonstration that the diversity is not fortuitous. It has not arisen from a whim or caprice of some deity. (Dobzhansky, *Genetics*, 2)

Well, this is not a book about religion, so I'll abstain from commenting on an unknown "deity." There is another way of looking at diversity, though. Certain individuals believe that life on Earth was created by an extraterrestrial civilization. Assuming that this is true, what would prevent extraterrestrials from introducing as much diversity as they want? A lack of patience, perhaps, but other than that there are no barriers. Dobzhansky tries to appeal to diversity as a proof of the evolutionary theory, but his proof is built on a shaky foundation.

## Peaks and Valleys

Geneticists introduced the notion of adaptive peaks and valleys in the field of gene combinations. Adaptive gene peaks are combinations of genes that make an organism most adaptable to its ecological niche; gene valleys, or unfavorable combinations of genes, make an organism unable to survive in any environment. Evolutionists were all too happy to adopt the idea without realizing that it disproves the evolutionary theory. This can be seen in the following scenario: suppose that ecological or environmental conditions have underdone changes, such changes being crucial to the production of new species (as the evolutionists have always maintained), and, as a result, an organism previously situated at the adaptive peak finds itself in the adaptive valley. Dobzhansky suggests that "the adaptive valleys are deserted and empty" (*Genetics and the Origin of Species*, 8), meaning that the organism is death-bound. Since all organisms, according to the evolutionists, went through adaptive valleys where no recovery was possible, the Earth by now should have been devoid of all organic forms as a dead planet! No one can destroy the evolutionary theory more thoroughly than the evolutionists themselves!

## Emotional Biases

Dobzhansky states that practically everyone accepts the evolutionary theory and that the exceptions are very rare and due to emotional biases.

> As recently as 1966, sheik Abd el Aziz bin Baz asked the king of Saudi Arabia to suppress a heresy that was spreading in his land.
>
> The sheik was perhaps unaware that the Space Age has begun before he asked the king to suppress the Copernican theory. The sphericity of the earth has been seen by astronauts and even by many earth-bound people on their television screens. Perhaps, the sheik could retort that those who venture beyond the confines of God's earth suffer hallucinations and that the earth is really flat. (Dobzhansky, "Nothing in Biology," 125–28)

Like any other evolutionist, Dobzhansky picks out a weak, or outright crazy, opponent and insists that all anti-evolutionists are stupid and emotionally unstable people, disregarding the fact that most of the damage to the evolutionary theory comes from geneticists, physicists, and mathematicians. Dobzhansky is not a very smart person; otherwise he would have attempted to counter a more serious criticism.

> If natural selection is the main factor that brings evolution about, any number of species is understandable: natural selection does not work according to foreordained plan, and species are produced not because they are needed for some purpose but simply because there is an environmental opportunity and genetic wherewithal to make them possible. Was the Creator in a jocular mood when he made *Psilopa petrolei* for California oil fields and species of *Drosophila* to live exclusively on some body-parts of certain land crabs on certain islands in the Caribbean? The organic diversity becomes, however, reasonable and understandable if the Creator created the living world not by caprice but by evolution propelled by natural selection. (Dobzhansy, "Nothing in Biology," 125–28)

If anyone is getting emotional here, that would be Dobzhansky.

What would prevent an extraterrestrial civilization from creating organic beings in accordance with the environmental conditions? Actually, this would be the only way to go; anything short of compliance with those conditions would spell death in a matter of hours. His remark about "jocular mood" is one of the strongest arguments that, from the evolutionist's standpoint, was put forward to defend the evolutionary theory. But this argument completely misses the mark because it is based on the unproven assertion that natural selection exists. From a logician's point of view, the existence of natural selection should be proved first, before any logical deduction is made.

Dobzhansky makes another highly emotional statement: there are scientists who already know the exact course of the evolutionary process. I wonder what paleontologists have to say about that? No one has claimed that the course is known or can be reconstructed, but Dobzhansky lets his emotions take over scientific judgment. But this is not all: Dobzhansky claims that other scientists have discovered the mechanisms that bring about evolution and can even run experiments to prove their theories. Well, if that were possible, we wouldn't be arguing about the evolutionary theory and all its opponents would be put to shame! It would be fair to say that Dobzhansky's passion and emotionality clouded his judgment.

## Genetics vs. Evolution

> Every succeeding generation of a species resembles but is never a replica of the preceding generation. Evolution is a development of dissimilarities between the ancestral and the descendant populations. (Dobzhansky, *Genetics*, 22)

Dobzhansky firmly believes that genetics proves the correctness of the evolutionary theory. If Dobzhansky's statement is correct, we would tend to expect genetic dissimilarity between parents and their offspring. This, however, is unheard of. In reality, genetic data proves that offspring include combinations of parental genes without the addition of any extraneous genetic material. Nothing provides stronger proof of the falsehood of the evolutionary theory than genetics.

Dobzhansky's poor knowledge of genetics becomes apparent when he discusses the experiment conducted by Beadle. During the experiment, certain biochemical reactions in the fungus *Neurospora* were blocked by what Dobzhansky believes to be a genetic mutation, causing an accumulation of cells that normally appear only at intermediate stages. Dobzhansky is unaware that geneticists often change biochemical reactions by the addition or removal of certain ingredients, or by changing the amounts and density of ingredients, for the purpose of better understanding the nature of complex biochemical reactions. But these changes do not qualify as mutations because they do not occur in a natural state, only in laboratory conditions. Dobzhansky's assertion is akin to saying, for example, that animal cloning involves some kind of mutation, which is not true because it takes place only in scientific laboratories.

Dobzhansky insists that the genetic rules that define individuals are different from the ones that define a population; he even offers an explanation of the differences. But a population is a group of individuals, so the distinction he makes is dubious. The implication is clear: it is impossible to change the genetic make-up of a population because its constituents always retain the same genetic structure. To support his assertion, Dobzhansky gives the following example: suppose certain environmental factors put either too tall or too short individuals at a disadvantage. Because of that the conditions will become lethal for tall (or short) individuals, and relative frequencies of homozygotes and heterozygotes in the growth genes will be changed.

If natural selection exists, it would change these frequencies to such a degree that there would be no tall or short individuals. Everybody would be roughly the same size, which has not been observed in any species so far.

Dobzhansky tries to define a Mendelian population, which is a hopeless endeavor:

> A Mendelian population is, then, a reproductive community of individuals which share a common gene pool.
> (Dobzhansky, *Genetics*, 79)

This is one of those definitions without the cause-and-effect relationship: is the community reproductive because they share a common gene pool, or do they share a common gene pool because the community is reproductive? Actually, this question does not have a clear answer, as is demonstrated by the following example. As demonstrated before, certain types of dogs do not interbreed even though by definition they share a common gene pool, and the wolf and German shepherd do interbreed even though they do not share the same gene pool (otherwise we would have to say that their offspring constituted a new dog-wolf category).

The lack of clarity in the definitions of genetic categories provides more proof of the falsehood of the evolutionary theory, which predicts, in disagreement with the nature, that there is a precise distinction between animals that belong to different genetic categories.

## Dobzhansky the Geneticist

Dobzhansky's articles on genetics are chock-full of incorrect and unworkable definitions. For example, he defines a genotype as the totality of genes of an individual and of a population at the same time. If these were true, all members of a population would have been genetic twins, like identical twin brothers or sisters. Yet Dobzhansky needs this idiotic definition to prove that changes, or mutations, in an individual's genotype cause changes in a population's genotype.

Dobzhansly states that the environment causes changes in a genotype—not the current environment, but the collective action of all historic environments. This could hardly be true. Geneticists deal with the current environment; the effects of past environments, even if they were observed in the past, cannot be modeled in laboratory conditions because there is never enough historic data to recreate them. Without the experimental proof, Dobzhanky's assessment falls apart; incidentally, this is also one of the breaking points of the evolutionary theory.

It is amazing to see how hopelessly confused Dobzhansky becomes when he talks about genotypes. First he claims that a genotype is "relatively quite stable" and then says that evolution is propelled by the interaction of an organism's genotype with the environment. What happened to the genotype's stability? If anything were unstable here, that would be Dobzhansky's knowledge of genetics! Of course, the genotype is stable, otherwise new species and subspecies would have been appearing every day.

How can the phenotype (the physical characteristics of the organism) be adaptive if it is controlled by the genotype, as Dobzhansky suggests? Dobzhansky's assessment of the dominant relationship between genotype and phenotype is correct, but the nature of this relationship implies that an adaptive or constantly changing phenotype cannot be guided by a relatively stable genotype. Perhaps Dobzhansky should not have ventured outside his primitive field, which is zoology; his misinterpretation of genetics makes him such an easy target.

Dobzhansky believes that the following mutually exclusive propositions are true: a) human intellectual and emotional development is shaped-up by the environment and b) traits of personality, such as reactions to the environment and intellectual capabilities, are determined by the genotype. A vast majority of specialists studying human behavior believe that neither one of the above propositions is true (in the author's opinion, proposition b is true, although small variations due to the environment are possible); a very few stupid ones think it is reasonable to posit the truth of two contradictory propositions. This may seem like a small error, but such errors stop the evolutionary theory dead in the tracks.

## Mutations and Phenocopies

By definition, a mutation transmits a newly acquired characteristic to the offspring while a phenocopy does not. Dobzhansky cites an interesting example of a mutation turning into a phenocopy when a certain bacterial organism's diet is changed. This transformation erases the boundary between mutations and phenocopies and raises an interesting question: what if such dietary changes that turn any mutation into a phenocopy really do exist? If the answer were yes, that would severely curtail the possibility of starting an evolutionary chain. In any case, this example demonstrates that the mutation mechanism is so poorly understood (and practically all geneticists will agree) that it simply makes no sense to base the evolutionary theory on the concept of mutation. Or, looking from a different perspective, it would be correct to say that the evolutionists grew so desperate that they included an unknown genetic mechanism in their theories.

## Lethal Mutations

In chapter II of *Genetics and the Origin of Species* Dobzhansky provides exclusive coverage of various kinds of mutation. It is interesting to note that almost all mutations are destructive or lethal, with very few being harmless, and none give any advantage to the organism. This shows that beneficial mutations are extremely rare, which comes as no surprise—mutations are caused by the destructive influences of X-ray radiation or chemical compounds that mess up genes. Now the whole evolutionary theory is in doubt because of the rarity of beneficial mutations. According to Dobzhansky, a number of mutations were needed to produce a new species; a single mutation would not suffice. This means that even if an organism is improved by one mutation, the other mutations are likely to be lethal, thus putting an end to any overall improvement.

All the mutations that Dobzhansky describes are induced, meaning that external factors, present either in the laboratory or in nature, cause them. But the evolutionists have been arguing that another type of mutation called "spontaneous mutations," which are allegedly responsible for the production of new species. Within this type of mutation are the following subdivisions: a) mutations occurring due to unknown factors and b) mutations occurring for no reason. Both definitions of spontaneous mutations are completely illogical. They pose serious questions without any answers: if the causes are unknown, how can anyone ascertain that they exist at all? It is utterly impossible to conduct an experiment based on unknown factors. If there are no causes, what happens to the cause-and-effect relationship? Not a single branch of science can survive if this relationship is removed from its foundation. The house of cards falls down, just like the evolutionary theory.

## The Hardy-Weinberg Law

This law is an algebraic equation that describes the genetic equilibrium within a population. It was discovered independently in 1908 by Wilhelm Weinberg, a German physician, and Godfrey Harold Hardy, a British mathematician.

The science of population genetics is based on this principle, which may be stated as follows: in a large, random-mating population, the proportion of dominant and recessive genes present tends to remain constant from generation to generation. In such a way even the rarest forms of genes, which one would assume would disappear, are preserved. (Encyclopedia Britannica, CD-ROM 2001)

Dobzhansky is happy to declare that this law is the foundation of the modern theory of evolution. All that remains to be said is that this is a very treacherous foundation; it shows better than anything else that the evolutionary theory completely contradicts reality, for the constancy of the ratio proves the impossibility of the creation of new species.

Actually, evolutionists say something else—they believe that the ratio does change because of mutations. But the law itself is based on the assumption that there are no spontaneous mutations, so the Hardy-Weinberg theorem in no way supports the evolutionary theory.

## Frequency of Mutations

From a mathematical standpoint, knowledge of the frequencies of naturally occurring or spontaneous mutations is crucial to the evolutionary theory. Without it it is impossible to tell whether evolution occurred. Dobzhansky spends a considerable amount of time and paper discussing the frequencies of induced mutations, but does not address spontaneous ones because this kind of data is unavailable. No one so far has observed a single spontaneous mutation, so that such mutations exist at all still remains a hypothesis.

Induced mutations produced in laboratory conditions are valuable tools in the investigation of complex biochemical reactions taking place in organisms. However, in no way do these conditions reflect the processes occurring in nature. Take, for example, mutations caused by X-ray radiation: large quantities of uranium ore are buried deep underground in a handful of places on Earth, yet they simply cannot cause a massive amount of mutations. Very few chemical compounds that cause mutations exist in natural form, and even the ones that do cannot by themselves break protective barriers and cause changes in genes. They have to be directly injected into a bio-structure to produce changes on a molecular level. The whole idea of looking at frequencies of induced mutations to infer any kind of statistical information about spontaneous mutations is absurd.

Despite the lack of evidence, Dobzhansky believes that mutants can be found in nature.

> Aberrant individuals found among masses of "normal" representatives of their species have often been recorded. Old-line naturalists classified such individuals as aberrations, phases, monstrosities, etc. In a number of cases it has been established that aberrations of this sort are in reality mutants, mostly recessive to the normal condition. They represent rare instances with recessive genes being carried in the population in heterozygous state, emerge as homozygotes because of the occasional mating of two carriers. (Dobzhansky, *Genetics*, 28)

Dobzhansky's explanation of "monstrosities" is correct, but these do not count as mutations because the recessive genes that resulted in such aberrations were already present in the gene pool. They add nothing new to the population genotype and, because of the absence of new characters, do not lead to the creation of a new species.

Besides, there is another cause of aberrations: unusually cold or hot weather often produces birth defects in insects.

## Genetics vs. Darwin's Principle of Natural Selection

Geneticists often use the concept of adaptive value, which is the frequency of transmission of a given combination of genes to the following generation. The Hardy-Weinberg theorem proves that the adaptive value remains constant for an indefinite period of time.

This is a valuable concept. However, it contradicts Darwin's principle of natural selection because organisms with the highest adaptive values are not necessarily the fittest ones. Dobzhansky acknowledges that "a genotype favored by differential survival at certain stages of the life cycle may have a lower net Darwinian fitness, if its superior viability is overbalanced by, for example, a lower fertility" (*Genetics and the Origin of Species*, 78), and at the same time insists that the adaptive value is the genetic equivalent of Darwinian fitness. Nothing could be further from the truth. Once again, Dobzhansky ended up contradicting himself. Ironically, one of the concepts most hailed by evolutionists renders the evolutionary theory a joke.

Dobzhansky believes that certain chemicals, including hydrocyanic gas and DDT, while destroying most organisms exposed to them, can produce beneficial mutations in a small percentage of individuals. However, there is no scientific data to support the hypothesis that these mutations actually occur. Most likely, the survived organisms were already immune to these chemicals because of their genetic structure and were able to pass on the "survivor gene" to the next generation. This is an artificial selection at work, but, as we noted before, it is not a natural, or Darwinian, selection.

Based on what they thought to be undisputed scientific data, Dobzhansky and several other evolutionists came to the conclusion that genetic changes causing alterations in one organ also affect developmentally related organs ("A mosaic of genes engenders an integrated development," Dobzhansky, *Genetics and the Origin of Species*, 108).

> Thus, relatively slight increase in body size results, because of heterogenic growth, in great increase of antler size in deer. (Huxley, "Problems," quoted in Dobzhansky, *Genetics*, 100)

> Establishment of genetic changes in one organ may require readjustments elsewhere in the body, through selection of mutations of other genes to restore a balanced genotype. (Reuch, "Historical changes," quoted in Dobzhansky, *Genetics*, 100)

> Reuch's statement is clearly nonsensical—how could a genotype be restored if a genetic change already occurred?

If an animal that weighs 300 kg suddenly balloons to 3 tons, it would need stronger, more massive legs to support its body, but, from a physical perspective, a slight increase in body size would not necessitate a dramatic change in antler size. The larger body would still be able to carry normal-sized antlers. If we assume that such changes occur, we come to the conclusion that the genome does not correspond to the physical structure of the body. This lack of correspondence implies that the organism cannot possibly survive in any environmental conditions. As often happens, Dobzhansky and other evolutionists misinterpreted experimental data. They needed this wacky theory to explain the presence of characters in organisms that are not essential to the survival of species (these neutral characters being contradictory to the evolutionary theory).

## Chromosomal Changes

Dobzhansky explains in some detail the formulas describing the frequency of changes of chromosomal types; Chapter V of his book is based on the works of R. Fisher, S. Wright, and others. The formulas are correct, but what relation do they have to the evolutionary theory? None whatsoever. These changes occurred either in laboratory conditions or due to seasonal weather patterns, which makes them cyclic, or bound to return to starting conditions. It is easy to guess that this cycle is detrimental to the evolutionary theory because a chromosomal change returns to the starting point without any progress being made in the creation of a new species. Dobzhansky tries to overcome this difficulty by saying "It is also a great, though highly misleading, simplification for a physiologist or a medical man to believe that different individuals, or patients, should react alike to similar treatments" (*Genetics and the Origin of Species*, 108). The analogy is incorrect; if this were true, a "medical man" would not be able to develop new prescription drugs. Because of his limited knowledge of mathematics, Dobzhansky does not understand how the Hardy-Weinberg theorem works. The theorem shows that when no changes are introduced into a population's genotype, the adaptive values remain the same.

## Human Races

In reference to human races, Dobzhansky notes that some anthropologists believe there are five races, others believe there are thirty, and so on. Despite this obvious disagreement, Dobzhansky states that racial diversity does not come as a result of arbitrary division but rather that different races constitute genetically distinct populations. The problem with his contention is that the meaning of the phrase "genetically distinct population" is not universally agreed upon. Different geneticists assign different meanings to it. For example, some believe that neo-Hawaiians and Asians are the same race while others do not. Everyone familiar with artificial intelligence methods knows that there are different definitions of the word *cluster*; the same is true in genetics where a race can be defined as a cluster with an arbitrary chosen unit of genetic distance between individuals. Apparently, Dobzhansky is unaware how clusters are categorized.

Dobzhansky remarks that there is no conclusive evidence to support the hypothesis that pigmentation of human skin depends on geographic location (this is one of the rare instances in which he is correct), but in the plant kingdom coloration of a plant depends on the climate. For example, tropical flowers have strikingly bright colors, while flowers growing in northern regions are pale. Well, someone may be interested to learn more about flowers and bees, but the origin of races remains an open question as far as the evolutionary theory is concerned. This is one of its most glaring deficiencies. Incidentally, there are several theories explaining racial differences, none of which is based on the evolutionary theory.

## Genetic Drift Revisited

R. Fisher has shown, with characteristic mathematic brilliance, that genetic drift practically disappears after fifteen generations (see chapter 13 of this book). S. Wright came to the diametrically opposed conclusion that genetic drift plays an extremely important role in the production of species. Dobzhansky, who sides with Wright, gives the following explanation: the Hardy-Weinberg law is applicable to populations of infinite size only; in real populations deviations from this law allow the production of new species to take place. He even provides some pathetic mathematical formulae developed by Wright to support this theory.

# Remark on the History of Turbulent
## Arguments between Wright and Fisher

Wright's study of variance, which led to the above assessment, was a source of "mathematical war" between him and American zoologists on the one hand and Fisher and British zoologists on the other. However, all professionals familiar with mathematical statistics, including this author, are on Fisher's side.

Now I'm going to provide a brief explanation of why Fisher is right. Everything hinges on the preciseness of the Hardy-Weinberg law. Yes, it was developed for an infinitely large population; this kind of approximation is used in mathematics and physics on a regular basis. All statistical approximations involving infinite populations give extremely good estimates for a population of one thousand and above; the error is practically negligible. For a population of one hundred the error is less than 0.1%; it becomes considerably larger when the population size drops below thirty. But such small isolated populations do not exist in nature. The whole story calls into question Wright's mathematical skill.

Just how crazy is Wright's mathematics? Dobzhansky provides the following example:

> If the entire progeny fails to survive, the mutant is lost; if one individual survives, the probability of loss of the mutant is 0.5; with two individuals surviving the probability of loss is 0.25; and with r survivors it is $2^{-r}$. (Dobzhansky, *Genetics*, 159)

Nobody counts probabilities like this: if one individual survives, the probability of loss of the mutant is zero, and then it remains zero no matter how many individuals survive. However, these numbers do not mean anything; instead, the probability of the survival of an individual, mutant or not, should be the basis of the calculations. But this is well beyond Wright's mathematical capabilities.

Dobzhansky provides another example of Wright's "abstruse mathematical argument":

> The smaller the effective population size, the greater are random variations in gene frequencies. (Dobzhansky, *Genetics*, 161)

No matter how Wright arrived at this conclusion, it contradicts all theoretical and experimental data that serves as the basis of statistics. If a large population is broken into smaller ones, the gene frequencies in some populations will exceed the average values in the large population and in others they will fall below, but if the grand average of these averages is calculated, its value will be practically equal to the average value in the initial, large population. Well, zoologists may believe in Wright's discoveries, but professionals with a mathematical background see them as garbage.

## Reproductive Isolation

Almost all evolutionists believe that some form of isolation is essential for the creation of new species. In chapter VII of his book Dobzhansky provides a lengthy discussion of various types of isolation. According to Dobzhansky there are two principal categories of isolation: a) geographic isolation and b) reproductive isolation, which, in turn, subdivides into ecological, sexual, mechanical, and other types of isolations.

I have already shown in the previous chapter that geographic isolation cannot possibly exist. Reproductive isolation is different in the sense that it does exist; however, not a single evolutionist has been able to explain how it developed. Unlike geographic isolation, which is supposedly imposed by external conditions, reproductive isolation is due to internal factors such as lack of male-female attraction, hybrid sterility, absence of correspondence of the genitalia, etc. If the evolutionary theory is correct, all these internal factors should stem from geographic isolation. Dobzhansky admits the lack of distinction between geographic and reproductive isolations by saying the following:

> The validity of distinction between these isolations has been questioned by some authors. Geographic isolation is on a different plane from all reproductive isolating mechanisms, because the former is independent of any genetic differences between populations, while the latter are necessarily genetic. (Dobzhansky, *Genetics*, 181)

Unless extraterrestrial scientists somehow installed the mechanism of reproductive isolation, evolutionists need to show how it emerged from nonexistent geographic isolation.

## The Shifting Balance Theory

In the final chapter of *Genetics and the Origin of Species*, Dobzhansky presents another of S. Wright's outlandish theories. Wright actually believed that he knew how evolution occurred.

> The theory itself has been controversial and confusing to many but has, at the same time, also helped many people to think about the evolutionary process in real populations. Ultimately, perhaps, the shifting balance theory has been not so much a rigorous theory, as a picturesque and thought provoking metaphor that has proven helpful in incorporating a spatial dimension into our thinking on evolution. (Joshi, "Shifting Balance," quoted at www.jncasr.ac.in)

In other words, other than nice little pictures of peaks and valleys in adaptive landscapes, Wright came up empty-handed as far as theoretical and experimental data is concerned. No other branch of science besides biology would seriously entertain a "scientific metaphor," even if it was thought-provoking. Nevertheless, let's take a brief look at this stillborn baby.

> The shifting balance theory of evolution was first laid out by Wright in two papers in 1931 and 1932. In this papers, Wright argued that the optimum situation for evolutionary advance, in the sense of a population becoming progressively better adapted to the environment, would be when a large population was divided and subdivided into partially isolated local races of small size. Wright illustrates these views through his concept of a field of gene combinations graded with respect to adaptive value, nowadays more commonly known as a fitness or adaptive landscape, or a surface of selective value. (Joshi, "Shifting Balance," quoted at www.jncasr.ac.in)

Now let's see what Wright himself has to say about his theory.

> One possibility is that a particular combination [of genes] gives maximum adaptive value and that the adaptivness of

the other combinations falls off more or less regularly. With something like $10^{1000}$ possibilities, it may be taken as certain that there will be an enormous number of widely separate harmonious combinations. The chance that a random combination is as adaptive as those characteristic of the species may be as low as $10^{-100}$ and still leave room for $10^{300}$ separate peaks, each surrounded by $10^{100}$ more or less similar combinations. (Wright, "Roles of Mutation," 356–66)

What if the chance that a random combination is as adaptive as those characteristic of the species is $10^{-999}$? Then Wright's theory breaks down! Which one is true? Without experimental data such determination is impossible. The experimental data itself should include the study of all environments, past and present, which would take about $10^{1000}$ years to study.

The elementary evolutionary process is, of course, change of gene frequency, a practically continuous process. Owing to the symmetry of the Mendelian mechanics, any gene frequency tends to remain constant in the absence of disturbing factors. If the gene mutates at a certain rate, its frequency tends to move to the left [on the graph of gene frequencies], but at a constantly decreasing rate. The gene type would ultimately be lost from the population if there were no opposing factor. But the gene type is in general favored by selection. Under selection, its frequency tends to move to the right. The rate is greatest at some point near the middle of the range. At a certain gene frequency the opposing pressures are equal and opposite, and at this point there is consequently equilibrium. (Wright, "Roles of Mutation," 356–66)

The phrase "certain rate" sounds almost scientific, but what exactly is such a rate? Wright did not provide a mathematical statistics theory that would prove the existence of such a rate.

Why under selection would the gene frequency tend to move to the right? There is no proof either. My calculations, stated in different terms (see the previous chapter of this book), show that the frequency moves to the left when natural selection is present, so the selection has an adverse effect on the production of new species.

It is amazing to see how Wright's wild and uneducated guesses became a "proven theory" for evolutionary biologists. Otherwise they would not quote him in their books.

## Another of Wright's Fiascos

This section is intended for professionals familiar with the methods of mathematical statistics.

This is almost a century-old argument—how to interpret Wright's path analysis.

> With the advent of Wright's method of path coefficients in early 20th century, the statistical landscape widened significantly. However, the adoption of Wright's methods provoked great controversy. Essential to this controversy, were claims, originally advanced by Wright, that this method could be applied to problems in which causality among variables could be assumed. (Denis and Legerski, "Causal Analysis," quoted at www.ingentaconnect.com)

One of the main goals of mathematical statistics is to determine whether dependency, or a causal relationship, exists between variables. Stating *a priori* that such a relationship exists is an extremely idiotic approach. It shows Wright's complete lack of knowledge of the discipline.

There are several methods to determine the relationships among variables. One of the best is R. Fisher's analysis of variance.

Proponents of path analysis insist that the method is correct because it closely resembles multiple regression. But multiple regression is not based on any assumptions about the nature of relationships among variables; it simply serves as a linear approximation of widely scattered experimental data and it is used to detect the trend. Sometimes this approximation is good, but in many cases, as in the case of stock price movements, it is far off the mark.

Just because Wright's formulae mirror multiple regression does not mean that they are correct; they are based on false premises and, because of that, do not serve the purpose and provide zero information about complex relations.

Fisher's analysis of variances enjoys broad usage and universal recognition; except for a handful of zoologists and botanists nobody has heard about Wright's path coefficients. Frankly, before I started writing this book I had not heard about Wright's works either. His obscurity is the best proof of the fact that Wright contributed nothing to the science of statistics.

## Fisher's View of Evolution

The reader may ask, if Fisher's works disprove the evolutionary theory so strongly, why, then, is his name listed among the names of the most prominent evolutionists? Fisher believed in natural selection as presented by Darwin, which is not a mathematical theory.

## Chain of Blunders

It all started with the faulty theory of path coefficients, which led to the creation of Wright's F-statistics and other silly stuff that, in turn, were used to "prove" that for small populations the Hardy-Weinberg theorem is incorrect, and then the conclusion that evolution is possible within small isolated populations followed.

All of Wright's legacy is a horrific mess of intertwined errors that only biologists put on the pedestal of science. The others see it for what it is. This is an example of pseudoscience putting on the mantle of true science.

## Quantum Mechanics & Random Mutations

Following Wright, many evolutionists believe that certain aspects of atomic or even nuclear motion can be used to explain the nature of random mutations. They often refer to Heisenberg's uncertainty principle as the cause of unpredictable changes on a macroscopic chromosome level.

Apparently, these "thinkers" have no idea how quantum mechanics works. To start with, it is impossible to develop a quantum theory based on the uncertainty principle that would show how a transition from the micro- to the macro- level could occur to account for a random mutation. For anyone familiar with quantum mechanics the reason is simple enough—a transition from the micro- to the macro-level is made for large assemblies of particles only when the average statistical properties of matter, such as the aggregate temperature, pressure, etc., are calculated on the basis of microscopic conditions, the average properties being observable on a macroscopic level. To predict the effect of the motion of a single particle such as an atom, electron, neutrino, etc., based on a macroscopic property of the matter one would have to measure its position and momentum simultaneously, but such measurement is prohibited by the uncertainty principle.

# 26 : Miscellaneous

## Punctuated Equilibrium

Supposedly, the incomplete fossil record was a source of constant pain for evolutionary biologists. It was only natural to expect someone to try to prove that evolution did not occur gradually, as Darwin proposed, but rather in jumps, without the development of intermediate forms. Therefore, a new theory called "punctuated equilibria" was proposed to account for the breaks in fossil records. It was created by N. Eldredge and S. J. Gould.

> Many breaks in the fossil record are real, they express the way in which evolution occurs, not the fragments of an imperfect record. The sharp break in a local column accurately records what happened in that area through time. (Eldredge and Gould, "Punctuated Equilibria," quoted at www.wikipedia.org)

The basic tenets of this theory were laid out in the article "Punctuated Equilibria: An Alternative to Phyletic Gradualism." This is a very strange article, to say the least. It begins with a quotation from P. B. Medawar:

> Innocent, unbiased observation is a myth. (Medawar, *Induction*, 28)

This idiotic proposition sets the tone for the whole article. To note in passing, this quotation suggests, for example, that the statement "at 0° C water turns into ice" is a biased, perhaps even wrong, observation. For some reason the authors feel compelled to quote Sir Isaac Newton—"I frame no hypotheses"—and then say that modern philosophers offered a great many explanations of this phrase. Apparently, the authors do not know how to interpret it either. Then what is the use of quoting it? (Actually, the interpretation is simple: "I offer proven scientific theories, not hypotheses.")

Things get even stranger after that:

Theory does not develop as a simple and logical extension of observation; it does not arise merely from the patient accumulation of facts. Thus, Hanson writes:

> Much recent philosophy of science has been dedicated to disclosing that a "given" or "pure" observation language is a myth-eaten fabric of philosophical fiction . . . In any observation statement the cloven hoof print of theory can readily be detected. (Hanson, *Perception and Discovery*, 22–23)

(Eldredge and Gould, "Punctuated Equilibria," quoted at www.wikipedia.org)

Hanson's statement is so ridiculous that it takes very little effort to disprove it: everyone familiar with the history of science knows that two kinds of electricity, negative and positive, were discovered when different materials were rubbed against each other. Prior to this discovery there was no theory describing such material properties. In fact, for a long period of time people didn't even know that electricity existed. Only after this discovery did Maxwell develop his theory of electromagnetic fields. Apparently, Hanson grossly misunderstood another statement: "A theory determines what kind of experiment to conduct." It means, among other things, if you develop a cosmological theory, you won't be testing it in a chemical lab.

If observation data loses its innocence and become tainted by theories, all objectivity is lost, but one still has to make a determination about which theory is true. It is a hopeless task for anyone, although Eldedredge and Gould claim it could be done.

> We wish to consider an alternative picture to phyletic gradualism; it is based on the theory of speciation that arises from the behavior, ecology and distribution of modern bio-species. First, we must emphasize that mechanism of speciation can be studied directly only with experimental and filed techniques applied to living organisms. No theory of evolutionary mechanism can be generated directly from paleontological data. Instead, theories developed by students of the modern biota generate predictions about the course of evolution in time. With these predictions, the paleontologist can approach the fossil record and ask the following question: Are observed patterns of geographic and stratigraphic distribution, and apparent rates and directions

of morphological change, consistent with the consequences of a particular theory of speciation? We can apply and test, but we cannot generate new mechanisms. If discrepancies are found between paleontological and expected patterns, we may be able to identify those aspects of a general theory that need improvement. But we cannot formulate these improvements ourselves. (Eldredge and Gould, "Punctuated Equilibria," quoted at www.wikipedia.org)

According to Eldredge and Gould, experimental data can be evaluated within the framework of a theory only; objective evaluation does not exist. This clearly implies that all scientists can claim that their incompatible theories are correct; there is no mechanism that would discriminate truth from falsehood. The fossil record itself can simply be interpreted the way a scientist thinks is appropriate, so his predictions always come true despite whatever patterns of geographic distribution he might observe.

Eldredge and Gould claim that their theory is the only one that predicts breaks in fossil records. This may be true among the evolutionist theories, but in general they are wrong. I can provide at least two theories that can account for the same thing: a) extraterrestrial scientists were using faulty genetic equipment incapable of producing continuity between intermediate forms and b) periodic volcanic activities produced so much hot lava that it destroyed the majority of intermediate species.

But the most damning evidence against the theory of punctuated equilibrium is genetic data—up until the present no sharp transitions were produced from one organism to another either in nature or in a genetic lab. All genetic theories clearly indicate that such transitions are impossible.

Maybe, there was some unknown factor in the past producing mutations of such magnitude, as the proponents of this theory would hope to be the case. That would be a huge factor doing such work for the last 2 billion years, and yet it has managed to go undetected. How great is the chance of that?

Now, let's compare 3 theories—a) punctuated equilibrium, b) faulty extraterrestrial equipment, and c) a multitude of volcanic eruptions—and ask ourselves, which one of them is the most plausible? So far no genetic evidence supporting either a) or b) has been found, but geological data shows that grand volcanoes existed in the past, which makes c) the most plausible theory out of these three. This does not mean that c) is correct, but it shows that c) beats a) and b) to the punch.

## Botany

Eldredge and Gould were not the only evolutionists troubled by the scarcity of information in fossil records. G. L. Stebbins tried another way out of the quandary by suggesting the following:

> Next comes the description of the differences between contemporary forms that are related to one another closely enough that the evolutionary process which gave rise to them can be inferred without reference to events that occurred in the remote past. (Stebbins, *Flowering Plants*, 3)

There are sciences that do not depend on past events, such as physical chemistry, solid state physics, dynamics of mechanical systems, etc. However, this is not true for all sciences. Take cosmology, for example. Astrophysicists rely heavily on data from the past in the form of background radiation to test their theories about the development of the universe. Without a theoretical reconstruction of past events cosmology is dead. The same is true for the evolutionary theory; without reference to past events it is completely useless. Unlike cosmology with its background radiation, the evolutionary theory does not have historic data to support itself.

### *Factors That Do Not Add Up*

Following Dobzhansky, Strebbins believes that five processes are responsible for evolution: mutation, genetic recombination, natural selection, chance fixation of genes, and reproductive isolation.

I already discussed four of these processes, and the one still remaining is the chance fixation of genes. It seems strange that someone would state that this process is different from random mutations; apparently, some evolutionists lack the necessary genetic background to see that no distinction exists. More importantly, mutations play very little, if any, role in the formation of new plants, as Stebbins admits.

> The review of mutations in flowering plants that have conspicuous effects on the phenotype suggests that the establishment of such mutations in population, though by no

means impossible, occurs so rarely that it must be regarded as exceptional rather than the usual basis for evolutionary trends. (Stebbins, *Flowering Plants*, 7)

Mutations are the only mechanisms that bring about changes in the gene pool. By admitting the rarity of mutations Stebbins in effect signed the death certificate of the evolutionary theory. It's no wonder that the other evolutionists (Dobzhansky, Mayr, Gould, etc.) were mute about the rarity of mutations.

## *His Majesty, the Experiment*

Stebbins is a unique figure among the evolutionists: he openly admits that theories based on the comparison between modern and fossil forms cannot be tested experimentally, while other evolutionists shy away from this topic. But what is a theory without experimental confirmation? It is just a useless interplay of words!

Instead of the experiment, Stebbins offers something else.

Indirect evidence must be of three kinds: 1) that which enables the evolutionist to infer the degree of similarity between the actual course of events at transspecific and subspecific levels; 2) that which permits the evolutionist to estimate the probability of the hypothesis that particular evolutionary trends have been guided and directed by the effects of natural selection; and 3) that which establishes the probability or improbability of alternative explanation, particularly internal direction of mutations. (Stebbins, *Flowering Plants*, 16)

Is it possible to define the "degree of similarity"? In certain sciences, including artificial intelligence, there are several definitions of this term and each one of them is defined in precise mathematical language. In biology and botany such definition is impossible, mainly because of the scarcity of fossil records, so the evolutionists are back to square one. The inability to provide even a vague definition of this term makes it devoid of scientific meaning; it becomes the equivalent of kindergarten talk.

The phrase "estimation of the probability of a hypothesis" has a precise mathematical definition. This concept is used in mathematical statistics on an everyday basis and such estimation involves precise numerical values. Stebbins offers nothing of this sort—there are no numerical values and not a single definition of the probability. He uses this expression in a metaphorical sense, which makes it unsuitable for a scientific conclusion.

## Terms with No Meaning

These terms with no meaning are "selective inertia," "conservation of organism," and "adaptive modification along the lines of least resistance" (in what units is this inertia measured? What are the lines of maximum resistance?). Stebbins uses them as if they had some quantitative meaning. In reality, these terms are not even metaphors because there are no points of reference; they are completely arbitrary. But such arbitrariness is the main characteristic of the evolutionary theory.

## Lack of Logic

Trying to explain plants' adaptations to various climatic conditions, Stubbins runs into a logical inconsistency.

> In the first place, unrelated evolutionary lines have often become adapted to similar environments, either via morphological and physiological convergence or by evolving different ways of exploiting the same environment. (Stebbins, *Flowering Plants*, 54)

First of all, how can evolutionary lines be unrelated if they all stem from the original cell? Common ancestry implies that even if a separation of lines occurred, they would all still be related. If they are related, why did they evolve in such different fashions? Morphological and physiological similarities in related plants suggest that they all should evolve in the same way, but this conclusion contradicts the observed data. The divergence of evolutionary lines suggests that certain plants did not evolve along the lines of least resistance, which clearly violates the principle of evolutionary canalization that Stebbins believes to be the basis of evolutionary development.

## Adaptive Trees

Stebbins uses trees thriving on soils with a high concentration of heavy metals as examples of natural selection at work. Unless we assume that all trees can live in such conditions, which is not the case, we must ask exactly how this natural selection occurred. Suppose a seed of a normal tree falls on a "heavy metal" soil. What would happen to it? Since it is not prepared to exist in such harsh conditions it will certainly die; a seed does not have any adaptive mechanism that would allow it to survive in conditions that it was not designed for. This example is one of the best proofs of the falsity of the evolutionary theory.

### Not-so-isolated Populations

Like all other evolutionists, Stebbins believes that some form of geographic isolation is needed for the production of a new species (this erroneous belief is based on Wright's perverse mathematical papers). There are three vehicles of cross-pollination: water (in aquatic plants), wind, and insects. Let's see what geographic isolation means in each of these three cases.

For sea plants, this form of isolation requires change in direction of the ocean currents, which has never been observed in the history of mankind (can the Gulf Stream change its course?). For river plants this would mean a reversal in the direction of the river's flow, which is a geological impossibility.

Can winds create an isolated area outside of which cross-pollination is prohibited? It could be done only if circular wind motion is sustained for a long period of time, which is a geophysical impossibility.

Could insects create a geographic isolation? If they were intelligent species they could, but otherwise only insect deaths can create this type of isolation. However, in this case there would be no future plant generations.

### From Cross-Pollination to Self-Fertilization

In chapter 4 of his book Stebbins provides examples of an alleged transition from cross-pollination to self-fertilization in several plants. How realistic is this transition? There is only one way to bring cross-pollination to an end—that would be the sudden, simultaneous deaths of the insect carriers. What is obvious to everyone, except for the evolutionists, is that if such death were to occur, it would leave the plants totally unprepared for the transition because nothing in their physiology shows that self-fertilization is an intrinsic characteristic. As a result, plants wouldn't be able to give rise to the next generation and, subsequently, the death of the species would follow.

There is another, deeper objection of philosophical nature to such transitions; it is based on the fact that adaptation to new conditions requires some form of intelligent thinking. For example, an animal predator encountered with a lack of prey in his original habitat is almost certain to decide to move into new geographic areas. Or an herbivore faced with a shortage of grass at his favorite pasture will likely decide to look for other places where the grass is greener. But no intelligent decision could be made on a tree's part, which shows that natural selection does not exist in the plant kingdom.

## Disagreement between the Theory and Observation

> The incidence of strong winds is erratic and unpredictable, so that they may or may not occur when seeds are ripe and ready to be transported. Birds and mammals, however, migrate at regular intervals, and so could exert a selective pressure on the time of ripening, favoring those plants that opened their seeds simultaneously with the animals' migrating habits. (Stebbins, *Flowering Plants*, 77)

Stebbins actually predicts the time when the seeds of certain plants are to open. If his predictions were correct they would add more credibility to the evolutionary theory. However, as direct observations show, there is no correlation between the times of animal migration and the openings of seeds. Somehow Stebbins managed not to mention this in his book.

## Unproven Theories in Genetics

> The evolutionist cannot trace directly the alterations of genes and gene-controlled processes that were responsible for evolutionary trends. He can observe only the outcomes of these changes in terms of alterations in the morphology and reactions of the adult organisms. (Stebbins, *Flowering Plants*, 107)

Once again, Stebbins admits what a majority of evolutionists are unwilling to talk about—that direct observation of changes in genetic structure cannot be done. But this puts the whole evolutionary theory in jeopardy because there is no proof that such changes occurred. As a rule, all sciences, except for biology, do not accept indirect proofs or inferences because there could be more than one interpretation of the outcomes. In the case of evolution one could argue that it did not take place at all; all genetic changes could have been programmed by extraterrestrial scientists, or perhaps there were no genetic changes but rather a variety of genetically similar organisms produced in the extraterrestrial lab.

> Recent research in developmental biology enables us to formulate some generalizations about the developmental pathways. (Stebbins, *Flowering Plants*, 110)

This is a gross logical error: developmental biology is based on the unproven hypothesis of developmental pathways, which, in turn, is inferred from developmental biology. As far as logic is concerned, this is a vicious circle.

Stebbins's poor knowledge of genetics resulted in the following formula:

> $A^n = a^m / a^i$ where $A^n$ is the number of similar organs or units, $a^m$ is the total number of meristematic cells capable of producing A-type parts, and $a^i$ is the number of meristematic cells needed to produce one A-type part.

First and foremost, Stebbins does not outline the criteria used to define similarity among the organs, which makes his formula completely useless. Because the similarity is not defined, it is impossible to tell what the right-hand part of the formula represents.

## Embryonic Similarity

Stebbins, like all other evolutionists, believes that similarity between the earlier stages of embryonic development of an organism and certain primitive, fully developed organisms proves that the evolution theory is correct. However, other explanations of this similarity that are not based on the evolutionary theory are also possible. One could argue, for example, that all mammals have certain characteristics common to each organism, these characteristics being developed during the gestation period. In fact, such characteristics do exist; they allow for the classification of a large number of species as mammalians. It does not follow, however, that these characteristics are the result of evolution.

## Poetic Justice

> Why cannot phylogenies, or even phylogenetic trees (evolution trees), be considered under the heading of poetry, or visual arts, or metaphors, or analogies, or hypotheses, all devices designed to lead us to a better understanding? To condemn a given phylogeny as speculative is as inappropriate as to damn a poet or composer for being imaginative. (Constance, "Systematic Botany," quoted in Stebbins, *Flowering Plants*, 115)

This is the best justification of the evolutionary theory that Stebbins has provided so far. Here is my response:

> Nor a beast, nor a mole, nor a single tree
> That spreads its branches beyond the hazy horizon
> In its wildest dreams that occur only
> When the Universe is asleep on its shoulders
> Saw the triumph of the Theory of Evolution.
> The Evolutionist's dream is dead, is dead, is dead!!

I just proved poetically that the evolutionary theory is false. Will the evolutionists accept this proof? I doubt it. But why should the world accept their ideas of poetic, evolved trees?

## An Evolutionary Noise Generator

In chapter 8 of his book Stebbins shares his thoughts on the ecological basis of plant diversity. Who would have thought that such small things as the direction of sun exposure, the steepness of slope, and depth of the soil would be capable of producing such diversity in plants! Well, perhaps, but Stebbins provides no observational data to support his guesswork, so the whole chapter is a waste of paper and of the reader's time.

In chapter 9 climatic influences on the evolutionary process are discussed. Once again, Stebbins makes plenty of noise, but provides no experimental data.

In chapter 10 a long table of angiosperm characters is presented. However, it is not clear how this data relates to the evolutionary theory. All angiosperms are contemporary; there is no historic data to show the changes, if there even are any. A creationist could have compiled this data; it does not prove either theory.

Trends of speciation within the angiosperms are described in chapter 11. It contains a lot of technical details that only botanists are interested in. It also could have been written by a monk or a member of a Canadian cult worshiping God the Extraterrestrial—either way it does not prove anything. I suspect that the last three chapters suffer from the same deficiency, but I did not have enough patience to read them.

## Paleontology

For the final review I choose the book *The Meaning of Evolution* written by G. G. Simpson. In this book Simpson tries to lay the philosophical foundations of the evolutionary theory.

As often happens with evolutionists, a gross misinterpretation of theoretical and experimental data is on display in chapter 2 of the book.

> Yet these studies show that there is no theoretical difficulty, under conditions that may well have existed early in the history of the earth, in the chance organization of a complex carbon-containing molecule capable of influencing or directing the synthesis of other units like itself. Such a unit would be, in barest essentials, alive. It would be similar or analogous to a virus. (Simpson, *Meaning*, 14)

In scientific parlance, the phrase "no theoretical difficulty" means that a theory already exists; in the case of the original cell that would mean that a theory existed explaining where it came from. But as everyone, including the evolutionists, knows, no such theory has been developed so far.

### Cambrian Explosion

Evolutionists define the Cambrian explosion as a sharp increase in the variety of organisms and transition form algae-like forms, or soft-bodied animals, to animals with hard body parts (shells, bones, etc.); this transition purportedly took place during the Cambrian period.

Simpson admits that it is hard to explain why simultaneous transitions in so many organisms occurred and offers no theories about how this might have happened other than to say that this is a fact. There is another inconsistency regarding the Cambrian explosion that Simpson prefers not to discuss —that not all animals acquired hard parts; a great many maintained their soft bodies. Naturally, the following question arises: what were the factors that caused a transition to solid bodies in some but not all soft-bodied animals? Apparently, these factors affected the Earth as a whole because solid and soft organisms are found in the same Cambrian geological formations; there is no geographic discrimination. It would be correct to say that these unknown factors were of cosmic magnitude, and yet nothing in the astronomic data suggests that they were present at any time in the past. Besides, why were the soft organisms affected in different ways? There is no answer to this fundamental question either, which puts the evolutionary theory at a disadvantage in comparison with the other theories. For instance, the proponents of the theory of alien intervention could say that both soft and hard animals were produced at the same time in extraterrestrial genetic labs as stages of grandiose genetic research. This contention explains, at least tentatively, the variety of forms, while the evolutionary theory is totally incapable of accounting for the Cambrian explosion. Although not universally accepted, there is a scientific principle according to which the weakest theories should be weeded out first. If this principle is to be followed, the evolutionary theory should be the first one to go.

## From Complexity to Simplicity

Evolutionists are known for making contradictory statements, often on the very same page, but none of them has topped Simpson. He managed to contradict himself in a single statement:

> Estimates of the number of individual animals stagger the imagination and are so very approximate as to be almost meaningless as figures, but it is safe to conclude that these numbers have increased markedly since the Cambrian. (Simpson, *Meaning*, 18)

This statement is meaningless but safe . . .

It appears that evolution goes both backwards and forwards; according to Simpson, when placed chronologically, organisms do not appear in the order of increased complexity. In other words, certain organisms had degraded to more primitive forms. This indicates that their adaptive qualities had been deteriorating, which contradicts the principle of natural selection. It is safe to conclude that Mother Nature is the strongest critic of the evolutionary theory.

### Rates of Evolution

Simpson gained notoriety among the evolutionists after introducing the concept of two evolutionary rates: the rate of structural changes and the rate of diversification (tempo and mode) for a given lineage. Being nonuniform, these rates depend on changing Earth conditions such as the formation of mountains and lakes, climatic development, volcanic activity, etc.

This seems to be a nice theory to test. However, there is no correlation whatsoever between the rate changes and geological cataclysms.

### Materialism and Vitalism

Referring to the definition of life (chapter 10), Simpson describes it as a competition between two philosophical schools, materialism and vitalism. Simpson's book was published in 1949. At that time positivism was already the dominant philosophical doctrine, as it is today. As a positivist, I find it strange that Simpson spends so much energy evaluating the sorely outdated doctrines of materialism and vitalism, doctrines that are of use to no one outside the field of biology. This example shows how incredibly outdated biological thought is.

### Most Likely to Be False

In chapter 11 Simpson provides examples of several conflicting theories, all of which, in his opinion, may be correct, although statistically speaking some of them are more probable than others.

This is something new that nonbiologists, including myself, have never heard about—the probabilistic assessment of theories. Statistical interpretation of data is a part of many scientific theories, but theories themselves are never judged on statistical grounds. To do so would imply that there are no absolute criteria telling whether a theory is correct or not. As all scientists know, excluding biologists, the absolute criterion of correctness of a theory is experimental data. Biologists believe that observational data can be evaluated within the framework of a theory only; data cannot be understood independently of some framework or other. But nonbiologists know that if this "rule" were really to be followed there would be no scientific progress. Simpson did not provide much philosophical discussion as far as scientific methodologies are concerned.

## *Bigger Is Better, but What about the "Small People"?*

> It is the most reasonable view, at least, that increase in gross size, which is among the most widespread of trends in animal evolution, is created by adaptation. (Simpson, *Meaning*, 138)

When Simpson wrote about this "evolutionary trend," the outside world knew nothing about the little people whose remains were recently discovered on a remote Indonesian island. Now this "trend" is left negated and the evolutionists have no idea about the direction of human evolution.

## *Hegel and Evolution*

> The word is only a convenient label for these tendencies in evolution: that what can happen usually does happen; changes occur as they may and not as would be hypothetically best. (Simpson, *Meaning*, 122)

It is almost as if Simpson is saying that what goes up must come down; at least this phrase would make sense. But another, meaningless phrase—that what can happen usually does happen—is a paraphrase of Hegel's philosophical maxim: The laws of nature are deterministic; nothing happens by chance; the mere presence of conditions necessary for the occurrence of an event guarantees that it will come to pass. Of course, this is not true, as everyone familiar with quantum mechanics knows. But biologists are firmly fixed in the past; for them Hegelian philosophy is as dominant as it was centuries ago. Perhaps their state of mind is predetermined by dead philosophical doctrines.

## Walking Blindfold

Occasionally, Simpson provides an interesting overview of competing evolutionist schools.

> Both schools submit that the image-forming eye could not function until *after* it was complete. One group, the final-ists, concludes that therefore the eye evolved with reference to a *future* function, that the structure began first and use for it came only millions of years later when it fully evolved ... They [another group] agree that since the image-forming eye, so they say, could not function until complete, therefore it became complete all at once. (Simpson, *Meaning*, 124)

Simpson himself believed that the eye developed in stages of increased complexity; at each stage it was capable of seeing.

Apparently, Simpson is familiar with neither physics nor physiology, otherwise he would have known that both groups are correct about the impossibility of the eye functioning before the final stage of its development. But then the finalists engage in some sort of unscientific mysticism, and their opponents believe in genetic miracles.

## Span of Life

> The potential span of life seems to be very rigidly set by the nature of the organism. All the great strides of medicine do not seem to have raised the potential life span of man by one minute. (Simpson, *Meaning*, 196)

Simpson is right about one thing—it directly follows from the evolutionary theory that the human life span cannot be raised substantially. This seemed to be a correct assessment in 1949 when Simpson's book was published. But the latest developments in prescription drug research and improvements in living standards have shown that this is not the case; the average life expectancy has been on the rise since the 1970s. This is the deadliest blow so far that the pharmaceutical industry has delivered to the evolutionary theory.

> The fact that extinction has not occurred for these animals during their exceptionally long histories does not permit the conclusion that their extinction never will occur. (Simpson, *Meaning*, 197)

Simpson moves to conclude that all species, including humans, must come to an end.

> The sun's energy is finite and must some day reach an end and with it life on the earth must cease. Collision with other celestial bodies is eventually probable in the endless time of forever. (Simpson, *Meaning*, 198)

This is another logical fallacy among the many that plague Simpson's book—no matter how long or short a species' lifespan has been, the mere fact of its longevity or youth is simply not sufficient to predict its fate, be that extinction or never-ending survival, and all positivists know that. But Simpson prefers to follow Hegel with his prediction that all things must come to an end. Once again, Simpson adopts a dead philosophical system.

As for the sun's energy being finite, there is plenty of time still before it burns out. Before that catastrophe happens man will have plenty of opportunities to either move to another planet outside the solar system or to reignite the sun.

## Simpson's View on Natural Selection

Simpson argues that criticism of Darwinian natural selection is no longer valid because the modern version of it is not based on intra- and extra-group struggle.

> To generalize from such incidents that natural selection in over-all and even in a figurative sense is the outcome of struggle is quite unjustified under the modern understanding of the process. Struggle is sometimes involved, but it usually is not, and when it is, it may even work against rather than towards natural selection. Advantage in differential reproduction is usually a peaceful process in which the concept of struggle is really irrelevant. It more often includes such things as better integration into the ecological situation, maintenance of balance of nature, more efficient utilization of available food, better care of the young, elimination of intra-group discords (struggles) that might hamper reproduction, exploitation of environmental possibilities that are not the objects of competition or are less effectively exploited by others. (Simpson, *Meaning*, 222)

Can a struggle work against natural selection? No one knows; Simpson did not provide a single example of such strange struggles.

Better integration into an ecological situation involves competition with other species, so some form of a struggle is inevitable.

"Maintenance of balance of nature" is such a broad phrase that it means practically nothing.

More efficient utilization of available food means direct competition and subsequent struggle, unless scavenging is involved, which is rare among species.

Intra-group discords do not hamper reproduction because sexual intercourse does not go non-stop for several days. There is plenty of time to make love and engage in a fight; these two activities do not interfere.

"Exploitation of environmental possibilities" is a paraphrase of "better integration into the ecological situation"—Simpson just repeats himself.

"Better care of the young" is a meaningless phrase because some species require less attention during the growth period than others; there is no common yardstick that could be used to nominate certain species as the most caring parents, and certain others as the worst parents.

Overall, Simpson failed to provide an alternative to the Darwinian struggle.

## Ethics and Man

It requires no demonstration that a demand for ethical standards is deeply ingrained in human psychology. Like so many human characteristics, indeed most of them, this trait is both innate and learned. Its basic mechanism is evidently part of our biological inheritance. Simpson, *Meaning*, 317)

According to the evolutionists, all ethnic groups share the same biological heritage in the ape. This clearly implies that the ethical code should be the same for all human societies. While murder is certainly not condoned in any culture, a number of religions such as Christianity, Buddhism, Judaism, Hinduism, etc., practice monogamy, while Muslims practice polygamy. It looks as though, in accordance with the evolutionary theory, Muslims have a unique biological inheritance; perhaps they even descended from a different ape, or perhaps a walrus. Unless we assume that ethical standards are imposed on civilizations from the outside, there is no possible explanation of this uniqueness.

In the animal world incest is not exceptional behavior, but in human societies it is strictly forbidden. Why does incest disagree with "our biological heritage"? There is no apparent reason for not having this type of sexual behavior unless an outside force imposed the prohibition. One reason that may be put forth is that incest leads to genetic degradation, but such deficiency could be easily overtaken if incest is combined with polygamy. For example, a man having a harem that includes his sister could issue an order that every time she gets pregnant the fetus should be aborted, while his other wives are allowed to give birth any time they like. Actually, this practice was widespread in ancient Egypt; it kept their genetic pool uncontaminated. Undoubtedly, the ancient Egyptians descended from a very different wacky ape.

Since the dawn of the evolutionary theory evolutionists of all persuasions have tried to understand how ethical standards were developed and what the connection is between evolution and ethics without realizing how deeply they have fallen into a logical fallacy. In order to show why they failed, I'll start with this question: is ethics based on certain considerations, whatever they may be? Unless we assume that ethics is completely arbitrary, there must be certain considerations upon which it is based. But what are those considerations? Unless we assume that they are arbitrary, they are based on other considerations, and so on, *ad infinitum*—the chain of logical inferences never ends. The only way out of this chain is to assume that ethics was not developed by any of Earth's civilizations, but was rather imposed by some outside organization, be that an extraterrestrial society or the Supreme Being.

# 27 : Scientific Realism

Out of all the modern philosophical systems, scientific realism is the only one that fully embraced the evolutionary theory. The others stayed away from this dicey topic. Its most prominent proponent, Karl Popper, incorporated the evolutionary theory into scientific realism. Popper also offered the following procedure of verification of scientific theories: suppose that theory B contains theory A as one of its integral parts; then if B is correct it would imply that A is also correct; similarly, if B is false then A is also false. Since scientific realism contains the evolutionary theory, the falsehood of this philosophical system extends to the concept of biological evolution. This is a good recipe for criticism of the evolutionary theory because scientific realism as a philosophical system is so weak that its falsehood can be easily seen.

This chapter is a bit different from the others. It does not include direct criticism of the evolutionary theory; instead, it is written for the purpose of dethroning scientific realism (despite its name, this is merely a pseudo-scientific theory).

## A Brief History of Scientific Realism

Scientific realism traces its origins to the ideas of the ancient Greek philosopher Plato, who believed in the reality of mathematical concepts in the sense that they represent attributes of nature existing independently of the human mind. This raises an interesting question that none of the realists can answer: if these attributes exist independently of the mind, then how was the mind able to learn about their existence? Positivists answered this question correctly when they pointed out that nothing exists outside the realm of perception.

A close connection between Plato's doctrine and scientific realism is discussed in detail in a very interesting book—*The Emperor's New Mind* by Roger Penrose. I'll note in passing that I fully agree with the basic premise of the book that computers cannot imitate the workings of the human mind. Penrose proved beyond the shadow of a doubt that there are inherent limitations that prevent computers from moving to a human-like stage. I recommend anyone interested in artificial intelligence or psychology to read this book. However, Penrose's view of quantum mechanics is completely wrong; apparently, the Copenhagen interpretation put forward by Niels Bohr is beyond his mental grasp. Penrose also shares a strange belief with many mathematicians that complex numbers somehow represent what the realists call "objective reality." However, there is no reality in numbers; complex or real, they are just concepts that physicists use to put their theories in observable form.

Other close predecessors of scientific realism are Kant and Hegel with their concept of transcendentalism. Hegel's idea of a quest for the universal scientific theory and a way of constructing it was adopted and further developed by Popper.

Realists go to great pains trying to distinguish themselves from idealists by stating that while idealists believe that the world consists of mind states only, they believe in the existence of a mind-independent world. From the positivist's point of view, no distinction can be made because both sides believe in the existence of unobservable entities. It would be correct to say that realism is a badly degraded form of idealism because, unlike realists, idealists see a direct connection between the mind and perception.

## Two Basic Tenets of Scientific Realism

1) It is possible to produce a universal scientific theory, the kind of theory that science aims to produce (Karl Popper described a methodology he believed to be essential to the creation of such a theory).

2) There exists a mind-independent world.

## Criticism of Scientific Realism

There are many angles of attack against scientific realism, but I choose to attack by way of the same principle upon which Popper based his methodology. Ironically, Popper did not realize that this principle could be used to obliterate scientific realism. Since Popper did not give the principle a name, I'll call it proposition Popper, or proposition P.

*Proposition P:* No matter how many times a process repeated itself in the past, it is incorrect to ascertain that it will repeat itself in the future.

Proposition P implies, for example, that no matter how many times the sunrise was observed in the past, there is no logical foundation for the assertion that it will continue happening in the future. As Popper correctly noted, the sun may explode tomorrow, in which case there would be no sunrise. Proposition P is correct, of course; even the lowly realists occasionally make valid statements. But Popper did not realize that proposition P also shows that it is impossible to produce a universal theory.

## The Impossibility of the Production of a Universal Theory

A majority of the modern cosmological theories lead to the conclusion that the universe has a finite mass. However, there are several theories predicting that the universe's mass is infinite. I am going to apply proposition P to both types of theories to show that it is impossible to produce a universal theory. I'll start with the infinite mass theories.

Clearly, a universe of infinite mass has an infinite size; otherwise we would have to assume that an infinite mass could be compressed to a finite size, which is absurd. Now, let's take a look at the physical laws in the vicinity of our sun. As scientists know, these laws are uniform, meaning that there are no variations in them. They are equally applicable to the Earth, Mars, Venus, etc. If we move to the stars in our metagalaxy, we can say with certainty that the laws will remain the same. However, the conclusion that these laws extend to the whole infinite universe is incorrect; proposition P ascertains that we cannot discard the possibility of having different laws of physics in outer regions of space. In fact, it is conceivable that an infinite universe consists of an infinite number of regions with different laws of physics. This means that a universal theory cannot be completed, which is another way of saying that it does not exist. And there is another problem with the idea of a universe of infinite mass: it would require an infinite number of differential equations to describe its evolution, which also means that a universal theory cannot be completed.

Now we turn to a universe of finite mass. Most likely it has a finite size as well. Such a universe could be divided into a finite number of regions with uniform laws of physics, so that would seem to provide a basis upon which to create a universal theory. However, we run into another difficulty—there is no guarantee that physical laws will never change. If proposition P is correctly applied, it leads to the conclusion that there is a distinct possibility of having a universe where the laws of physics are constantly, albeit very slowly, changing. Once again, we come to the conclusion that a universal theory cannot be completed.

There is another way of looking at things—none of the modern theories predicts the creation of a universal theory. The reason is simple: being parts of a universal theory, all other theories are less comprehensive than the theory that they comprise; they are unable to step out of their boundaries and ascertain that they are components of a bigger thing. The only theory capable of predicting the production of a universal theory is the theory of a universal theory itself. But this is a logical contradiction: since the universal theory does not exist, it cannot predict its own creation! Therefore, it is impossible in principle to provide a proof of the existence of a universal theory.

Realists are certain to argue that no one so far has been able to prove that it is impossible to create a universal theory. But this is not the point. What we know now is that scientific realism is based on a premise that cannot be classified as either true or false, which is a logical error of monumental proportions.

## *Pathetic Doctrine*

Every natural science begins with a set of postulates. For example, in physics the postulates are Newton's laws, Maxwell's electromagnetic equations, etc. Every postulate is supposed to be confirmed by experimental data. For instance, Einstein's postulate that the speed of light is independent of the velocity with which a system of coordinates moves has a direct experimental verification. The truth or falsity of a postulate is determined by the experiment; at least this is how positivists see scientific development.

The realist's version of science includes postulates whose truth or falsity cannot be verified by way of experiment because they refer to what realists call "real objects," which lie beyond the sphere of perception. The only means of accessing these objects is by way of logical analysis. But their assertion leads to an unresolvable logical contradiction, as I will show.

Suppose we have two postulates, A and B, whose truth or falsity cannot be verified by way of experiment. How would we know that A does not contradict B? The only way to draw a conclusion regarding the relation between A and B is to perform a logical analysis. But how can we guarantee that a logical analysis is correct? The only way to do that is to perform an analysis of the analysis. Then again, how can we guarantee that the analysis of the analysis is correct? And so on, the chain of analyses goes on *ad infinitum*, which means that it is impossible to draw any concrete conclusion regarding the relationship between A and B. This also means that scientific realism makes no distinction between true and false statements.

## Conclusion

Scientific realism falls apart in the presence of even the slightest criticism. It took me only two paragraphs to destroy this pseudoscientific doctrine; in contrast, it took a whole book to destroy the theory of evolution. In somewhat poetic terms, scientific realism espoused the evolution theory, and it happened to be true love till the death.

# Epilogue

The history of mankind is fraught with wrong theories—astrology, alchemy, psychoanalysis, and ether, to name a few. However, all these theories have one thing in common: they were all initiated as a quest for knowledge, and then something went terribly wrong. The evolutionary theory is different in this respect: it started out as a scientific fraud and has continued as such. Darwin purposely distorted scientific data in order to perpetrate his racist views and become famous in the process. Darwin wanted to be remembered as an even more famous scientist than another Englishman, Sir Isaac Newton. Newton's fame was like a poison in Darwin's morning cup of tea.

Evolutionists will certainly argue that Charles Darwin was not the first scientist to propose the evolutionary theory. Erasmus Darwin and others proposed it long before it acquired its final form. But prior to Darwin's failed research the evolutionary theory was understood to be merely a hypothesis, not a theory. Darwin fraudulently raised it to a status that it doesn't deserve.

As mankind progresses and true sciences develop, all fraudulent and incorrect theories disappear into a sort of cosmic repository of wrong ideas. The time has come for the evolutionary theory to go away; the cosmic bin is eagerly waiting.

# Appendix A

This appendix is intended for readers familiar with the probability theory. This is a simple case where the average number of cubs for an "animal family" is one.

Denote by A an event of the parent transmitting certain characteristics to an offspring; by B an event of transmitting this characteristic to the "grandchild."

Denote a survival rate for the species by n.

In probability terms, $P(A) = n$; also $P(B|A) = n$; (this is conditional probability).

Find $P(B)$.

Step 1)   $P(B|A) = P(AB)/P(A)$   $P(A)=P(A|B)=n \rightarrow P(AB)=n^2$

Step 2)   $P(A|B) =1$ This means that the occurrence of event B indicates that event A has already happened. Also, $P(A|B)=P(AB)/P(B)$  from step 1) $P(AB)=n^2 \rightarrow P(B)=n^2$

It is easy to show, by using mathematical induction, that the probability of transmitting a characteristic to the k-th generation equals $n^K$.

# Appendix B

The purpose of this article is to show that the laws of nature rule out any process that would lead to the creation of the original cell.

In general, all existing matter could be divided into two groups:

Group A—materials that do not possess DNA-like structure.

Group B—materials that possess DNA-like structure.

Materials that belong to Group B contain information about themselves, which means that they can duplicate themselves, while materials belonging to Group A do not contain information about themselves, which means that they *cannot* duplicate themselves. Materials in Group B contain information about themselves in their DNA.

The word *information* in this context designates all biochemical processes that occur in a chemical compound. The phrase "self-reproductive information" will also be used in this article to designate information that allows a material to duplicate itself. Clearly, only materials from Group B possess self-reproductive information.

The following is a well-known scientific fact: materials from Group B can be decomposed into their basic ingredients (the basic ingredients are the elements that form the periodic table); however, no one has succeeded in putting those basic ingredients back together again once they've been broken apart such that the original material from Group B is restored.

What kind of information does a basic ingredient, or an element from the periodic table, possess? It possesses information about the number of its electrons, protons, neutrons, their energy levels, the probability of transition between the levels, and so on. If we have two basic ingredients that form a chemical bond with each other, we can deduce the properties of their compound from their individual properties. In a vast majority of the cases, deduction of the properties of a compound is an extremely difficult task and, because of the amount of mathematical calculations, it has been done in only a handful of cases. But such a deduction is, in principle, possible.

What happens to the self-reproductive information when a material from Group B is decomposed into its basic ingredients?

> **Proposition I.** This information is stored in the basic ingredients or in the set of rules that govern interactions between basic ingredients.

Does a basic ingredient possess self-reproductive information? No, it doesn't; otherwise, it would be able to duplicate itself. Materials from Group B, on the other hand, possess self-reproductive information as well as the information about their basic ingredients. Although basic ingredients do not possess self-reproductive information, according to Proposition I they possess reproductive information about more complex structures of which they are a part. Therefore, Proposition I leads to the following contradictions: a) According to definition, materials that possess such information belong to Group B. Basic ingredients belong to Group A. This means that Group A contains more information about Group B than it does about itself. From a methodological standpoint, this means that classification is impossible. In other words, it is impossible to determine whether the DNA-like structure is a reality. b) Not a single element of the periodic table contains information about any element other than itself—this is a scientific fact. If this were not true, there would be no difference between the intrinsic and extrinsic properties.

Could this information be obtained from the set of rules that govern interactions between basic ingredients? But these rules simply represent the properties that are common to all basic ingredients; they do not contain any new information (these rules are put in place for the reason of convenience only). Therefore, there is only one possible conclusion: Proposition I is incorrect.

> **Proposition II.** This information is stored elsewhere.

Perhaps it is stored in some kind of cosmic repository or in anthropologists' heads, or perhaps it was miraculously transformed into background cosmic radiation, or perhaps Mother Earth keeps it as one of her deep secrets—all of these are purely metaphysical speculations that the evolutionists may be willing to put forward.

**Proposition III.** This information is destroyed in the process of decomposition of a material from Group B.

The proof is very simple—this is the only remaining option.

In practical terms, what does the word *destroyed* mean in this case? It means that, since the information is gone, it is impossible to restore a material from Group B to its original form from its basic ingredients.

This impossibility of restoration is another way of saying that the creation of a Group-B material from scratch, or from basic ingredients, is impossible. This also means that a process that would lead to the creation of the original cell can't possibly exist.

# Bibliography

Conze, Edward. *Buddhist Scriptures.* New York: Penguin, 1959.

Darwin, Charles. *The Descent of Man.* 2nd ed. New York: A. L. Burt, 1874.

———. *The Origin of Species.* Mentor Edition. New York: The New American Library, Inc., 1958.

Denis, Daniel J., and Joanna Legerski. "Causal Modeling and the Origins of Path Analysis." *Theory & Science,* 2006.

Constance, Lincoln. "Systematic Botany—an Unending Synthesis," *Taxon* (1964) 257–73.

Dobzhansky, Theodosius. *Evolution, Genetics, and Man.* Wiley, 1955.

———. *Genetics and the Origin of Species.* 3rd ed. New York: Columbia University Press, 1951.

———. "Nothing in Biology Makes Sense Except in the Light of Evolution." *The American Biology Teacher* (1973) 125–28.

Eldredge, N., and S. Jay Gould. "Punctuated equilibria: an alternative to phyletic gradualism." In *Models in Paleobiology,* edited by Thomas J. M. Schopf, 82–115. San Francisco: Freeman, Cooper, 1972.

*Encyclopedia Britannica.* 2001. CD-ROM.

Hanson, Norwood. *Perception and Discovery; an Introduction to Scientific Inquiry.* San Francisco: Freeman, Cooper, 1970.

Huxley, J. S. *Problems of relative growth.* London: Methuen, 1932.

Joshi, Amitabh. "The Shifting Balance Theory of Evolution." *Resonance* (Dec. 1999) 66–75.

Mayr, Ernst. *Systematics and the Origin of Species, from the Viewpoint of a Zoologist.* Cambridge: Harvard University Press, 1999.

Medawar, Peter. *Induction and Intuition in Scientific Thought.* Philadelphia: American Philosophical Society, 1969.

"Mutation, Gene Flow, and Recombination. Drift and Selection." In Botany online - The Internet Hypertextbook of the University of Hamburg. http://www.biologie.uni-hamburg.de/b-online/e40/40d.htm.

Reusch, B. "Historical changes correlated with evolutionary changes of body size." *Hereditas.* 1948.

Simpson, George. *The Meaning of Evolution.* Third Edition. New Haven: Yale University Press, 1960.

Stebbins, G. Ledyard. *Flowering Plants: Evolution Above the Species Level.* Cambridge: Belknap Press of Harvard University Press, 1974.

Wright, Sewall. "The Roles of Mutation, Inbreeding, Crossbreeding and Selection in Evolution." *Proceedings, 6th International Congress of Genetics* (1932) 356–66.

———. "The Material Basis of Evolution," *The Scientific Monthly* 53 (1941) 165–70.